METHODS IN MOLECULAR BIOLOGY

Series Editor
John M. Walker
School of Life Sciences
University of Hertfordshire
Hatfield, Hertfordshire, AL10 9AB, UK

For further volumes:
http://www.springer.com/series/7651

Mass Spectrometry Imaging of Small Molecules

Edited by

Lin He

Department of Chemistry, North Carolina State University, Raleigh, NC, USA

 Humana Press

Editor
Lin He
Department of Chemistry
North Carolina State University
Raleigh, NC, USA

ISSN 1064-3745   ISSN 1940-6029 (electronic)
ISBN 978-1-4939-1356-5   ISBN 978-1-4939-1357-2 (eBook)
DOI 10.1007/978-1-4939-1357-2
Springer New York Heidelberg Dordrecht London

Library of Congress Control Number: 2014953846

Printed on acid-free paper

Humana Press is a brand of Springer
Springer is part of Springer Science+Business Media (www.springer.com)

Preface

Rapid technological advances have been made in recent years to address challenges facing researchers in the life science field, especially in the studies of complex mechanistic pathways within biochemical and biological organisms. Among them, Mass Spectroscopy Imaging (MSI) has emerged as an enabling technique to provide insight into the molecular entities within cells, tissues, and whole-body samples and to understand inherent complexities within biological metabolomes. In this Springer Protocols volume, *Mass Spectrometry Imaging of Small Molecules: Methods in Molecular Biology*, experts in the MSI field present cradle-to-grave protocols for two-dimensional, and three-dimensional in some cases, visualization and quantification of a wide array of small molecular species present in biologically relevant samples.

This book is meant as a practical guide to provide operational instructions, from sample preparation to method selection, from comparative quantification to structural identification, from data collection to visualization, for small molecule mapping in complex samples. Our aim in this volume is to bring the rapidly maturing methods of metabolic imaging to life science researchers and to minimize technical intimidation in adapting new technological platforms in biological research.

Following a brief presentation of the technique background in Chapter 1, the content of the book is arranged according to different ionization methods used, one of the key factors in MSI to ensure effective and efficient metabolite imaging. The book starts with two most widely used ionization methods for small molecule imaging: secondary ion mass spectrometry (SIMS) (Chapters 2 and 3) and matrix-assisted laser desorption and ionization methods (MALDI) (Chapters 4–6). Chapters 7–13 address a common problem encountered in analyzing tissue samples in vacuum and provide solutions of modern ambient ionization methods based on direct electrospray and/or laser ablation. Using nanomaterials in place of organic matrices, Chapters 14–17 describe a group of surface-assisted laser desorption and ionization methods using various inorganic matrices/structures to reduce background noises in the low mass region. Chapter 18 presents the use of a customized instrument for selective ionization of analytes of interest with post-photoionization. It offers better imaging sensitivity and reduced fragmentation in some cases, attractive features regardless of the absence of commercial instrumentation. While almost all chapters touch on the issues of data processing and analysis, Chapter 19 is dedicated to address challenges and options to be considered in data handling and provides a set of generic procedures to be followed in data visualization and statistical analysis.

I would like to thank all writers for their willingness to share their knowledge of the field and their time and efforts to contribute. I would also like to thank the publisher, Springer, and Emeritus Professor John M. Walker at University of Hertfordshire, who not only encouraged me to take on this writing project but also provided invaluable guidance along the way to make this book a reality.

There is an old Chinese saying: 工欲善其事,必先利其器. (One must have good tools in order to do a good job.) I hope this book will bring you a set of good tools to tackle biological challenges at hand and pursue exciting new research trajectories the heart sets.

Raleigh, NC, USA   *Lin He*

Contents

Contributors

ARTEM AKHMETOV • *Department of Chemistry, University of Illinois at Chicago, Chicago, IL, USA*

RACHEL V. BENNETT • *School of Chemistry and Biochemistry, Georgia Institute of Technology, Atlanta, GA, USA*

CHHAVI BHARDWAJ • *Department of Chemistry, University of Illinois at Chicago, Chicago, IL, USA*

ANNA BLOOM • *Department of Chemistry, Penn State University, University Park, PA, USA*

VICTORIA L. BROWN • *Department of Chemistry, North Carolina State University, Raleigh, NC, USA*

ALAIN BRUNELLE • *Centre de Recherche de Gif, Institut de Chimie des Substances Naturelles, CNRS, Gif-sur-Yvette, France*

ELAINE C. CABRAL • *Department of Chemistry, Faculty of Science, Centre for Research in Mass Spectrometry (CRMS), York University, Toronto, ON, Canada*

ADAM D. FEENSTRA • *Department of Chemistry, Iowa State University of Science and Technology, Ames, IA, USA; U. S. Department of Energy, Ames Laboratory, Ames, IA, USA*

FACUNDO M. FERNÁNDEZ • *School of Chemistry and Biochemistry, Georgia Institute of Technology, Atlanta, GA, USA*

ERIN GEMPERLINE • *Department of Chemistry, University of Wisconsin—Madison, Madison, WI, USA*

LUKE HANLEY • *Department of Chemistry, University of Illinois at Chicago, Chicago, IL, USA*

LIN HE • *Department of Chemistry, North Carolina State University, Raleigh, NC, USA*

MIN-ZONG HUANG • *Department of Chemistry, National Sun Yat-Sen University, Kaohsiung, Taiwan*

DEMIAN R. IFA • *Department of Chemistry, Faculty of Science, Centre for Research in Mass Spectrometry (CRMS), York University, Toronto, ON, Canada*

CHRISTIAN JANFELT • *Section for Analytical Biosciences, Department of Pharmacy, University of Copenhagen, Copenhagen, Denmark*

SIOU-SIAN JHANG • *Department of Chemistry, National Sun Yat-Sen University, Kaohsiung, Taiwan*

TAKESHI KONDO • *Department of Cell Biology and Anatomy, Hamamatsu University School of Medicine, Hamamatsu-shi, Shizuoka, Japan*

ANDREW R. KORTE • *Department of Chemistry, Iowa State University of Science and Technology, Ames, IA, USA; U. S. Department of Energy, Ames Laboratory, Ames, IA, USA*

MICHAEL E. KURCZY • *Department of Chemistry, Molecular and Computational Biology, Center for Metabolomics and Mass Spectrometry, The Scripps Research Institute, La Jolla, CA, USA*

INGELA LANEKOFF • *Physical Sciences Division, Pacific Northwest National Laboratory, Richland, WA, USA*

JULIA LASKIN • *Physical Sciences Division, Pacific Northwest National Laboratory, Richland, WA, USA*

YOUNG JIN LEE • *Department of Chemistry, Iowa State University of Science and Technology, Ames, IA, USA; U. S. Department of Energy, Ames Laboratory, Ames, IA, USA*

HANG LI • *Department of Chemistry, W. M. Keck Institute for Proteomics Technology and Applications, The George Washington University, Washington, DC, USA*

LINGJUN LI • *Department of Chemistry, School of Pharmacy, University of Wisconsin—Madison, Madison, WI, USA*

JIANGJIANG LIU • *Weldon School of Biomedical Engineering, Purdue University, West Lafayette, IN, USA*

QIANG LIU • *Department of Chemistry, North Carolina State University, Raleigh, NC, USA*

LÁSZLÓ MÁRK • *Department of Analytical Biochemistry, Institute of Biochemistry and Medical Chemistry, University of Pécs, Pécs, Hungary*

TARA N. MOENING • *Department of Chemistry, North Carolina State University, Raleigh, NC, USA*

KERMIT K. MURRAY • *Department of Chemistry, Louisiana State University, Baton Rouge, LA, USA*

TRENT R. NORTHEN • *Life Sciences Division, Lawrence Berkeley National Laboratory, Berkeley, CA, USA*

ZHENG OUYANG • *Weldon School of Biomedical Engineering, Purdue University, West Lafayette, IN, USA*

SUNG-GUN PARK • *Department of Chemistry, Louisiana State University, Baton Rouge, LA, USA*

KEIGO SANO • *Department of Cell Biology and Anatomy, Hamamatsu University School of Medicine, Hamamatsu-shi, Shizuoka, Japan*

MITSUTOSHI SETOU • *Department of Cell Biology and Anatomy, Hamamatsu University School of Medicine, Hamamatsu-shi, Shizuoka, Japan*

JENTAIE SHIEA • *Department of Chemistry, National Sun Yat-Sen University, Kaohsiung, Taiwan*

BINDESH SHRESTHA • *Department of Chemistry, W. M. Keck Institute for Proteomics Technology and Applications, The George Washington University, Washington, DC, USA*

GARY SIUZDAK • *Department of Chemistry, Molecular and Computational Biology, Center for Metabolomics and Mass Spectrometry, The Scripps Research Institute, La Jolla, CA, USA*

BRIAN K. SMITH • *Department of Chemistry, W. M. Keck Institute for Proteomics Technology and Applications, The George Washington University, Washington, DC, USA*

EIJI SUGIYAMA • *Department of Cell Biology and Anatomy, Hamamatsu University School of Medicine, Hamamatsu-shi, Shizuoka, Japan*

DAVID TOUBOUL • *Centre de Recherche de Gif, Institut de Chimie des Substances Naturelles, CNRS, Gif-sur-Yvette, France*

SUNIA A. TRAUGER • *Center for Systems Biology, Harvard University, Cambridge, MA, USA*

AKOS VERTES • *Department of Chemistry, W. M. Keck Institute for Proteomics Technology and Applications, The George Washington University, Washington, DC, USA*

MICHIHIKO WAKI • *Department of Cell Biology and Anatomy, Hamamatsu University School of Medicine, Hamamatsu-shi, Shizuoka, Japan*

NICHOLAS WINOGRAD • *Department of Chemistry, Penn State University, University Park, PA, USA*

XINGCHUANG XIONG • *Weldon School of Biomedical Engineering, Purdue University, West Lafayette, IN, USA*

GARGEY B. YAGNIK • *Department of Chemistry, Iowa State University of Science and Technology, Ames, IA, USA; U. S. Department of Energy, Ames Laboratory, Ames, IA, USA*

Chapter 1

Current Status and Future Prospects of Mass Spectrometry Imaging of Small Molecules

Victoria L. Brown and Lin He

Abstract

In the field of small-molecule studies, vast efforts have been put forth in order to comprehensively characterize and quantify metabolites formed from complex mechanistic pathways within biochemical and biological organisms. Many technologies and methodologies have been developed to aid understanding of the inherent complexities within biological metabolomes. Specifically, mass spectroscopy imaging (MSI) has emerged as a foundational technique in gaining insight into the molecular entities within cells, tissues, and whole-body samples. In this chapter we provide a brief overview of major technical components involved in MSI, including topics such as sample preparation, analyte ionization, ion detection, and data analysis. Emerging applications are briefly summarized as well, but details will be presented in the following chapters.

Key words Mass spectroscopy imaging (MSI), Metabolite, Sample preparation, MS ionization, MS analyzer

Metabolites are the class of low-molecular-weight molecules resultant from metabolic processes within biological systems. The quantitative and qualitative studies of metabolites are designed to enhance our existing knowledge of the metabolome, the intricate compilation of metabolites within cells, tissues, biofluids, and organs found in eukaryotic organisms. Characterization of global response patterns, metabolic pathways, and cell-specific functions in plant and animal metabolites provides insight into cellular and physiological changes in response to genetic or environmental stressors and is of great importance in the fields of agriculture, pharmacology, and molecular medicine [1–4].

Ever since the bloom of metabolomics studies, mass spectrometry (MS) has been the method of choice for the field, owing to its unprecedented resolving power and sensitivity [5, 6]. Providing snapshots of spatial distribution of biologically relevant metabolites in complex samples, mass spectrometry imaging (MSI) provides additional dimension of information to metabolite profiling hence it enables direct 2D visualization, and subsequent

Lin He (ed.), *Mass Spectrometry Imaging of Small Molecules*, Methods in Molecular Biology, vol. 1203, DOI 10.1007/978-1-4939-1357-2_1, © Springer Science+Business Media New York 2015

correlation, of their concerted regulatory roles in metabolic network [7–10]. Strategic fingerprinting each chemical component based on its unique molecular composition, MSI has become the workhorse technique in gathering heterogenic topological information from biological samples to elucidate metabolite pathways associated with various biochemical processes that are multifarious in nature.

In a typical MSI experiment setting, samples are canvassed in a raster of points, each of which has a corresponding x and y coordinates (sometimes with the axis z monitored as well). As the sampling beam scans over the sample, an independent mass spectrum is collected for the predefined mass range. Spectra for each raster point, or x–y coordinate, are then processed and reconstructed into a 2D (or 3D) molecular image of the system being analyzed. Achieving universal detection and quantification of small molecules simultaneously, however, is a nontrivial task. A robust workflow with carefully designed sample preparation, data management, and quantification strategies is a must. First and foremost, regardless of the sample types, properly handling and preparation are critical to ensure preservation of the molecular morphology and chemical integrity for reproducible and reliable results [11–13]. For example, physical (low temperature) or chemical (fixation reagents) means have been used to suppress metabolite degradation. Cleverly designed and carefully executed strategies have been developed to process natural samples of great variations (e.g., patient to patient, diet, age, disease state), which are reiterated with great emphasis throughout this book.

Nonbiased liberation of analytes from a solid surface into gas phase, with well-preserved spatial locations and relative concentrations, for subsequent MS analysis is the next critical step in MSI experiments. Breakthroughs and improvements in various desorption and ionization techniques have significantly broadened the repertoire of ionization sources for MSI. Common ionization methods used in MSI include secondary ion mass spectrometry (SIMS), matrix-assisted laser/desorption ionization (MALDI), desorption electrospray ionization (DESI), and their derivatives, all providing a multitude of advantageous features in their own way [7, 8].

Ionization through SIMS entails a high-energy primary ion beam to generate secondary ions from a surface [14–16]. It has long been used for elemental analysis of thin films and materials because the ion beam can be easily focused in an electric field with submicron resolution. Successful extension to small-molecule profiling in cellular and tissue imaging has been demonstrated along with the new generation of ion beams (e.g., C_{60}^+ or Au cluster ion beam) and has become one of the routine approaches in subcellular 2-D and 3-D imaging [16].

The acclaim-to-fame of MALDI is its effective ionization of large molecules, along with electrospray ionization, that changes the landscape of proteomics studies. Its applications to small-molecule analysis was hindered initially by the presence of high matrix background but it becomes less of an issue in recent years with rapid advancement of high-resolution mass analyzers [17–19]. For MALDI and other ionization sources that involve matrix-coated samples, choosing an appropriate matrix is vital for a successful measurement [20, 21]. Subsequent matrix deposition is another important step to ensure reproducible matrix coating and uniform crystal formation, which has direct effects on limiting analyte migration and preserving spatial resolution and ion intensities [20–23]. Techniques, both manual and automated, have been created to achieve optimum reproducibility of matrix deposition for MSI experiments. Successful examples include dried-droplet, pneumatic nebulization, sublimation, spray-coating, acoustic deposition, ChIP, electrospray, solvent-free, and chemical inkjet printing [24–29]. While these approaches share the same goal, each differs in the size of matrix crystals formed, economic feasibility, practicality, and throughput, so a thorough inspection of resulting samples is necessary in order to choose the most appropriate method for a given MSI experiment. Despite the arduous nature of confining each component in matrix deposition, utilizing MALDI-MSI has explicit advantages for metabolite imaging: its versatility has been shown in the diverse subjects studied, such as whole-body samples, cancerous tissue diagnostics, and drug development [6, 7, 30, 31].

Development of ambient-based MSI techniques offers a more effective solution to the issues associated with sample stability under vacuum. In this regard, DESI has established itself as a versatile and dynamic ambient ionization source with considerable capabilities for metabolite imaging [13, 32]. As one of the most widely used ionization methods nowadays, DESI has minimized potential matrix interference in small-molecule detection [32–34] and has been successfully utilized in imaging whole-body sections, profiling compounds in forensic studies, detecting antifungal metabolites within natural products, monitoring metabolic species in plant imprints, and characterizing varying disease states within tissue samples. While the large scanning footprint due to the size of the incoming charged droplet beam is one of the trade-off of DESI, various improvements to reduce beam size and better ionization selectivity, as discussed in later chapters, have made great strides in high-resolution MS imaging [13, 35, 36].

Taking advantages of nanotechnology development, novel nanomaterials have been incorporated in various means for analyte ionization with better ionization efficiency and minimized sample preparation [37–40]. Desorption/ionization on silicon (DIOS) was the most successful one based on porous silicon substrate. The more effective version of DIOS by facilitating the direct and unbiased detection and imaging of metabolites within biofluids and biological tissues [41, 42]. Substrates such as porous metal oxides, metal nanoparticles, carbon nanotubes, and thin films have all been implemented as viable surface pretreatment options, either individually or in conjunction with a matrix-based application, to optimize results in MSI experiments.

Needless to say, although a plethora of ionization methods have been developed and are presented in this protocol book, selection of the appropriate method for specific samples is decided by desired imaging results, as each ionization source possesses its own intrinsic benefits and limitations in the achievable spatial resolution, mass range and sensitivity, etc. This book includes a handful of examples on how different ionization methods are used in analyzing similar samples and vice versa, and more importantly, the considerations took place behind such decisions.

Upon successful releasing of analytes into gas phase, an additional element is required for any sound and successful metabolite imaging experiment—an efficient mass analyzer [6, 43]. In the broadest sense, mass analyzers function as a high precision balance by differentiating molecules and molecular fragments based on their mass-to-charge (m/z) ratios. The most widely used ones are time-of-flight (TOF), Fourier transform ion cyclotron resonance (FTICR), and Orbitrap mass analyzers. Time-of-flight (TOF) mass analyzers detect ions based on their kinetic energy in either a linear or reflectron mode. The latter mode uses electrostatic mirrors to adjust for slight differences in kinetic energy after the desorption/ionization event, which leads to a pronounced improvement in mass resolving power [43]. Tandem MS analyzers in the form of TOF–TOF provide additional capability for pertinent structural elucidation of analytes through extensive fragmentation. Together with its high throughput, superior sensitivity, and large m/z range, TOFs are the predominant mass analyzers used in MSI experiments.

Other mass analyzers such as Fourier transform ion cyclotron resonance (FTICR) and ion-trap/orbitrap have become more and more popular for their high mass accuracy, imperative features for ambiguous distinction of minuscule mass differences [6, 43]. FTICR is much different from TOF mass analyzers as detection of analytes is based on the frequencies of the ions traveling in orbits. FTICR is advantageous for metabolite imaging purposes as it offers high mass measurement accuracy (MMA), high resolving power, and multiplexed MS/MS capabilities, but at the price of lower throughput when compared to TOF [44, 45]. The applications, but due to their low mass accuracy they are frequently coupled to another mass analyzer to enhance performance [46, 47]. Orbitrap is another modern MS analyzer that provides excellent mass resolving power and mass accuracy, comparable to FTICR. In a simplistic sense, orbitraps function by confining ions around a central electrode. The axial oscillations of these ions are translated into an image current, where the m/z ratios of the ions can then be determined. Orbitraps are most often coupled to linear ion traps (LITs) mass analyzers for general metabolomic studies orbitraps to achieve fast scan speeds and sensitive detection, despite low ion abundance levels [46–49].

While there exist powerful mass analyzers and a myriad of ways to combine them into dynamic tandem systems, the common goal between each piece of technology is the same—to deliver the best quality results that encompass molecular specificity, highly resolved spectra, and reliable, reproducible data [22, 50–53]. Therefore, instruments well equipped with state-of-the-art ionization sources and mass analyzers only paint a part of the picture—the data acquired in an imaging experiment can be massive in size and tremendously time consuming. The sheer size of the data files to be processed imposes great challenges to decipher between biologically significant data and basal background. In order to combat the challenges that accompany acquired experimental data and to mitigate the difficulties in handling such elaborate data sets, many software-based and statistical methods have been developed and applied to MSI experiments to help quantitatively and qualitatively process, evaluate, and analyze metabolomic information. Some of these methods include but are not limited to data preprocessing (e.g., baseline subtraction, normalization, spectral recalibration, spectral averaging, and spectral intensity threshold adjustments) and various forms of multivariate statistical analysis. While many chapters in this book primarily use tools provided by instrument manufacturers for data visualization and analysis, Chapter 19 discusses direct comparison of different commercial software. For more detailed discussion on the topic, extensive reviews and book chapters are available elsewhere.

There is no doubt that MSI has emerged rapidly as an indispensable tool for metabolomics as the information collected composes core components of systems biology stands complementary to other "-omics" fields. The approach is straightforward with minimal chemical modification in order to carry out analysis and does not require a priori knowledge of the sample. While it is important to note that MS in general is an invasive and destructive technique, with strategic selection of an appropriate sample preparation method, ionization means, and mass analyzer, successful mapping of small-molecules in subcellular compartments, tissues, and/or whole-body samples has nevertheless been widely practiced in daily metabolomic research. Rapid advancements in MSI instrumentation and better sample processing methodologies, in combination with development of powerful data analysis tools and biological database, will indisputably further expand its capability and invite more researchers in life science to explore its potentials in metabolomic studies.

References

1. Fiehn O (2002) Plant Mol Biol 48:155
2. Kell DB, Goodacre, R (2014) Metabolomics and systems pharmacology: why and how to model the human metabolic network for drug discovery. Drug Discovery Today 19:171
3. Zenobi R (2013) Science 342:1201
4. Liesenfeld DB, Habermann N, Owen RW, Scalbert A, Ulrich CM (2013) Cancer Epidemiol Biomarkers Prev 22:2182
5. Kell DB, Brown M, Davey HM, Dunn WB, Spasic I, Oliver SG (2005) Nat Rev Micro 3:557
6. Ibanez C, Garcia-Canas V, Valdes A, Simo C (2013) Trends Anal Chem 52:100
7. Lietz CB, Gemperline E, Li L (2013) J Adv Drug Deliv Rev 65:1074
8. Miura D, Fujimura Y, Wariishi H (2012) J Proteomics 75:5052
9. Griffiths WJ, Koal T, Wang Y, Kohl M, Enot DP, Deigner H-P (2010) Angew Chem Int Ed 49:5426
10. Soares DP, Law M (2009) Clin Radiol 64:12
11. Solon E, Schweitzer A, Stoeckli M, Prideaux B (2010) AAPS J 12:11
12. Amstalden van Hove ER, Smith DF, Heeren RMA (2010) J Chromatogra A. 1217:3946
13. Wiseman JM, Ifa DR, Zhu Y, Kissinger CB, Manicke NE, Kissinger PT, Cooks RG (2008) Proc Natl Acad Sci 105:18120
14. Ostrowski SG, Van Bell CT, Winograd N, Ewing AG (2004) Science 305:71
15. Brunelle A, Laprevote O (2009) Anal Bioanal Chem 393:31
16. Passarelli MK, Ewing AG, Winograd N (2013) Anal Chem 85:2231
17. Li Y, Shrestha B, Vertes A (2007) Anal Chem 79:523
18. Nordstrom A, Want E, Northen T, Lehtio J, Siuzdak G (2008) Anal Chem 80:421
19. Shroff R, Rulisek L, Doubsky J, Svatos A (2009) Proc Natl Acad Sci U S A 106:10092
20. Caprioli RM, Farmer TB, Gile J (1997) Anal Chem 69:4751
21. Greer T, Sturm R, Li L (2011) J Proteomics 74:2617
22. Schwartz SA, Caprioli RM (2010) Imaging mass spectrometry: viewing the future. Methods Mol Biol 656:3
23. Garrett TJ, Prieto-Conaway MC, Kovtoun V, Bui H, Izgarian N, Stafford G, Yost RA (2007) Int J Mass Spectrom 260:166
24. Kaletaş BK, van der Wiel IM, Stauber J, Lennard JD, Güzel C, Kros JM, Luider TM, Heeren RMA (2009) Proteomics 9:2622
25. Sugiura Y, Setou M, Horigome D, Setou M (ed) (2010) Springer, Japan, p 71
26. Svatos A (2010) Trends Biotechnol 28:425
27. Hankin J, Barkley R, Murphy R (2007) J Am Soc Mass Spectrom 18:1646
28. Aerni H-R, Cornett DS, Caprioli RM (2006) Anal Chem 78:827
29. Trimpin S, Herath TN, Inutan ED, Wager-Miller J, Kowalski P, Claude E, Walker JM, Mackie K (2010) Anal Chem 82:359
30. Stoeckli M, Staab D, Schweitzer A (2007) Int J Mass Spectrom 260:195
31. Vrkoslav V, Muck A, Cvacka J, Svatos A (2010) J Am Soc Mass Spectrom 21:220
32. Dill AL, Eberlin LS, Ifa DR, Cooks RG (2011) Chem Commun 47:2741
33. MuÃàller T, Oradu S, Ifa DR, Cooks RG, KraÃàutler B (2011) Anal Chem 83:5754
34. Thunig J, Hansen SH, Janfelt C (2011) Anal Chem 83:3256
35. Laskin J, Heath BS, Roach PJ, Cazares L, Semmes O (2012) J Anal Chem 84:141
36. Nemes P, Barton AA, Li Y, Vertes A (2008) Anal Chem 80:4575
37. Thomas JJ, Shen Z, Crowell JE, Finn MG, Siuzdak G (2001) Proc Natl Acad Sci 98:4932
38. Greving MP, Patti GJ, Siuzdak G (2011) Anal Chem 83:2
39. Liu Q, Xiao Y, Pagan-Miranda C, Chiu Y, He L (2009) J Am Soc Mass Spectrom 20:80
40. Liu Q, He L (2009) J Am Soc Mass Spectrom 20:2229
41. Lee DY, Platt V, Bowen B, Louie K, Canaria CA, McMurray CT, Northen T (2012) Integr Biol 4:693
42. Patti GJ, Shriver LP, Wassif CA, Woo HK, Uritboonthai W, Apon J, Manchester M, Porter FD, Siuzdak G (2010) Neuroscience 170:858
43. Chughtai K, Heeren RMA (2010) Chem Rev 110:3237

44. Bedair M, Sumner LW (2008) TrAC Trend Anal Chem 27:238

45. Cornett DS, Frappier SL, Caprioli RM (2008) Anal Chem 80:5648

46. Douglas DJ, Frank AJ, Mao D (2005) Mass Spectrom Rev 24:1

47. Aebersold R, Mann M (2003) Nature 422:198

48. Peterman SM, Duczak N Jr, Kalgutkar AS, Lame ME, Soglia JR (2006) J Am Soc Mass Spectrom 17:363

49. Manicke NE, Dill AL, Ifa DR, Cooks RG (2010) J Mass Spectrom 45:223

50. Norris JL, Cornett DS, Mobley JA, Andersson M, Seeley EH, Chaurand P, Caprioli RM (2007) Int J Mass Spectrom 260:212

51. Hendriks MMWB, Eeuwijk FAV, Jellema RH, Westerhuis JA, Reijmers TH, Hoefsloot HCJ, Smilde AK (2011) Trends Anal Chem 30:1685

52. Tong D, Boocock D, Coveney C, Saif J, Gomez S, Querol S, Rees R, Ball G (2011) Clin Proteom 8:14

53. McDonnell LA, Heeren RMA (2007) Mass Spectrom Rev 26:606

Chapter 2

Sample Preparation for 3D SIMS Chemical Imaging of Cells

Nicholas Winograd and Anna Bloom

Abstract

Time-of-flight secondary ion mass spectrometry (ToF-SIMS) is an emerging technique for the characterization of biological systems. With the development of novel ion sources such as cluster ion beams, ionization efficiency has been increased, allowing for greater amounts of information to be obtained from the sample of interest. This enables the plotting of the distribution of chemical compounds against position with submicrometer resolution, yielding a chemical map of the material. In addition, by combining imaging with molecular depth profiling, a complete 3-dimensional rendering of the object is possible. The study of single biological cells presents significant challenges due to the fundamental complexity associated with any biological material. Sample preparation is of critical importance in controlling this complexity, owing to the fragile nature of biological cells and to the need to characterize them in their native state, free of chemical or physical changes. Here, we describe the four most widely used sample preparation methods for cellular imaging using ToF-SIMS, and provide guidance for data collection and analysis procedures.

Key words Time-of-flight (ToF), Secondary ion mass spectrometry (SIMS), Freeze-fracture, Freeze dry, Frozen hydrate, Chemical fixation, Single cell

1 Introduction

Chemical imaging on the cellular level is now feasible with time-of-flight secondary ion mass spectrometry (ToF-SIMS). The ability to map the distribution of small biologically relevant molecules (<1,000 Da) is now possible due to the submicron spatial resolution, high chemical specificity, and high surface sensitivity of ToF-SIMS [1, 2]. In brief, a pulsed high-energy primary ion beam is used to bombard the surface. As the beam rasters across a sample surface, material is ablated in the form of neutrals and ions; these are the secondary ions that will be analyzed in the mass spectrometer. These secondary ions are extracted into a ToF analyzer from which a mass spectrum is created. Each pixel contains a full mass spectrum corresponding to a specific position of the beam.

Lin He (ed.), *Mass Spectrometry Imaging of Small Molecules*, Methods in Molecular Biology, vol. 1203, DOI 10.1007/978-1-4939-1357-2_2, © Springer Science+Business Media New York 2015

Hence, a 2D image can be created by selecting a specific mass and plotting its intensity pixel by pixel [3].

Due to the ultrahigh-vacuum conditions typically required for a ToF-SIMS experiment, special precautions must be taken to conserve sample integrity when analyzing cells. In order to obtain mass spectral data with sufficient quality, a surface must be chosen on which cells will successfully grow, any interfering chemicals or components must be removed, and the shape and distribution of chemical components in the cell must be maintained [4, 5]. Generally, cells must be fixed either cryogenically or chemically, and so commonly used sample preparation methods include chemical fixation, freeze fracturing, frozen hydration, and freeze-drying. We outline these four sample preparation protocols, with brief descriptions given below.

Chemical fixation preserves the internal cellular structures by physically changing the chemistry of the cell using reagents. Several chemical compounds are commonly used, the most common of which are glutaraldehyde (GA), paraformalin/formaldehyde, and trehalose [6–8]. A major advantage of this technique is that samples can be analyzed at room temperature, rather than using a cryogenically cooled stage, as is the case with many other cell preparation methods. This approach preserves the integrity of the cellular compartments containing the chemical information that is generally of interest; however, the integrity of the cellular membrane is often compromised and the distribution of diffusible ions is not retained during chemical fixation [6–9].

For sample preparation techniques utilizing freezing, the sample is generally flash-frozen in liquid nitrogen-cooled propane prior to freezing in liquid nitrogen. This ensures that the cellular and tissue structures sustain little or no damage due to water crystallization, subsequently, maintaining cellular integrity [10].

Freeze-drying involves quickly freezing the sample in order to preserve the chemistry, followed by slow warming under vacuum to remove residual water [3]. Cell rupturing is often a major concern, due to the sublimation of water if temperatures of the sample are increased too quickly. It has also been shown that freeze-drying can cause rearrangement of molecules within the cell, sometimes even causing certain components to be lost completely [4, 6, 11]. This makes freeze-drying less than ideal when it is the exact location and distribution of chemicals and components in the cell that are of interest [1, 6, 11, 12].

In freeze fracturing, the cell suspension is trapped between two shards of substrate in a sandwich format and subsequently plunge frozen in liquid propane (88 K). Cells prepared in this way are stored in liquid nitrogen until analysis can be performed. Moderate force is then applied and the ice matrix within the sample plates is fractured, exposing the cells contained within. Freeze fracturing has been shown to maintain chemical heterogeneity in the cell, allowing for accurate chemical analysis, but causes a fracture plane that is not always reproducible, making intact cellular characterization, particularly on the surface, difficult [11–13].

Frozen hydration is believed to be best suited to preserving the integrity of the cell and has been shown to increase ion yields for some species [14, 15]. Here, cells remain in their native hydrated state, thus minimizing the risk of chemical movement or dissipation while still cryogenically fixing cells to prevent problems associated with analysis in vacuum. However, it poses the greatest challenge in terms of sample handling. The sample must remain at cryogenic temperatures after flash-freezing in order to remain in the frozen hydrated state, meaning that instrumentation must be equipped with cold-stage capabilities [1, 7, 11, 12].

In addition to various fixing methods, thought must be given to extraneous compounds that may be in contact with the cell. A major problem occurs with the salt residue remaining on the cells after culturing, as this can be detrimental to the SIMS signal. Berman et al. were able to determine that washes in ammonium formate remove these residual salt peaks from the buffer solution without physically damaging or chemically altering the cell [4, 6, 12].

When determining which cellular sample preparation protocol to follow, it is up to the researcher to determine which method is best suited to the experiment, as each of the preparation methods described above highlights a slightly different type of information. A protocol for each of these approaches is outlined in Subheading 3. These protocols have been optimized to ensure compatibility with the SIMS experiment.

2 Materials

The various sample preparation methods utilized various cell types and cell culture protocols. The reader is asked to consult the specific references for more information regarding cell culturing. Sample preparation processes will be described beginning with the grown cells unless otherwise noted.

2.1 Freeze Fracture [5]

1. HeLa cells, an immortalized cervical cancer cell line, were used in this experimental protocol.

2. Poly-L-lysine: 0.01 % solution.

3. Steel shards.

4. Ammonium formate: 0.15 M.

5. Deionized water.

6. Propane gas.

7. Liquid nitrogen.

8. Instrumentation: J105-3D Chemical Imager (Ionoptika Ltd., UK) equipped with a 40 keV C_{60}^+ primary ion source.

2.2 Frozen Hydrate [13]

1. HeLa cells, an immortalized cervical cancer cell line, were used in this experimental protocol.

2. HPLC-grade hexane.

3. Ethane gas (99 %).

4. Liquid nitrogen.

5. Ammonium formate: 0.15 M.

6. Deionized water.

7. 5×5 mm silicon shards (Ted Pella, Redding, CA, USA).

8. Instrumentation: Bio-ToF (Ionoptika Ltd., UK) equipped with 40 keV C_{60}^+ primary ion source [16].

2.3 Chemical Fixation with Glutaraldehyde [6]

1. Infinity telomerase-immortalized primary human fibroblasts (hTERT-BJ1) (Clontech Laboratories, Inc., Mountain View, CA) were used in this protocol.

2. Polished silicon wafers, 1 mm × 1 mm.

3. Phosphate-buffered saline (PBS; 0.02 M NaH_2PO_4, 0.02 M, Na_2HPO_4, 0.15 M NaCl, 5.4 mM KCL, pH 7.2).

4. Glutaraldehyde in PBS: 2.5 % solution.

5. Instrumentation: TOF-SIMS IV (ION-TOF GmbH, Germany) equipped with 25 keV Bi_3^+ primary ion source.

2.4 Chemical Fixation with Paraformalin/ Formaldehyde [7]

1. HeLa M cells, an immortalized cell line derived from cervical cancer, were used in this protocol.

2. Poly-L-lysine: 0.01 % solution.

3. Silicon shards.

4. Dulbecco's modified Eagle medium.

5. Phosphate-buffered saline.

6. Formalin: 4 % solution.

7. Millipore water.

8. Ammonium formate: 0.15 M.

9. Instrumentation: J105 3D Chemical Imager (Ionoptika, Ltd., Southampton, UK, and SAI Ltd., Manchester, UK) equipped with 40 keV C_{60}^+ primary ion beam.

2.5 Chemical Fixation with Trehalose [8]

1. J774, murine, peritoneal macrophages from 4-week-old CBA/J male mice and glial cells from 1-day-old Sprague-Dawley rat pups were used for this protocol.

2. Glass shards: Shards are coated with <100 Å Cr and <100 Å Au, followed by soaking in poly-L-lysine (1.2 mg/mL solution) and collagen (0.1 mg/mL solution), rinsing in deionized water, and air-drying.

3. α-α (d) trehalose: 50 mM solution.

4. Phosphate-buffered saline (PBS) washing solution: PBS containing 50 mM trehalose and 10–15 wt% glycerol.

5. 200-mesh finder grids (Electron Microscopy Sciences).

6. Instrumentation: Described previously [16] equipped with 5–25 kV Au primary ion source (Ionoptika).

2.6 Freeze-Dry [17]

1. Xenopus oocytes (Professor Mark Boyett, University of Manchester) were used for this experimental protocol.

2. Propane gas.

3. Liquid nitrogen.

4. Ammonium acetate: 0.15 M.

5. Deionized water.

6. 5 × 5 mm silicon shards (Ted Pella, Redding, CA, USA).

7. Instrumentation: BioToF-SIMS (Ionoptika Ltd. UK) equipped with a 40 keV C_{60}^+ primary ion source and a 25 keV liquid metal ion gun (LMIG) fitted with Au:Ge eutectic source (Ionoptika Ltd.) providing Au^+ and Au_3^+ ions.

3 Methods

As described in Subheading 1, the methods for freeze fracture, frozen hydrate, freeze-dry, and chemical fixation cellular sample preparation will be outlined here. While these methods may not be the only possible cellular sample preparation methods, they have been found to be the most successful in maintaining the chemical integrity of the cell for analysis. The protocols described have been utilized with specific cell types; however, analysis is not limited to these particular cell types. It is assumed that the reader has some familiarity with ToF-SIMS analysis.

3.1 Cell Culture

1. Culture the cells of choice in the manner described by the protocol for each specific type.

2. Prior to growing cells, shards must be cleaned. The process for silicon shards includes placing the desired number of shards in a glass scintillation vial and covering with nanopure water. The water is removed and methanol is added until the shards are covered. Shards are sonicated uncovered for 5–10 min. The methanol is removed and the shards are dried with a gentle stream of nitrogen gas. This process is repeated with heptane and acetone. Store the shards in methanol (*see* **Notes 1** and **2**).

3.2 Freeze Fracture [5]

1. Grow cells on poly-L-lysine-coated steel, a hinged two-plate substrate specifically designed to fracture biological samples in mousetrap device (*see* **Note 3**).

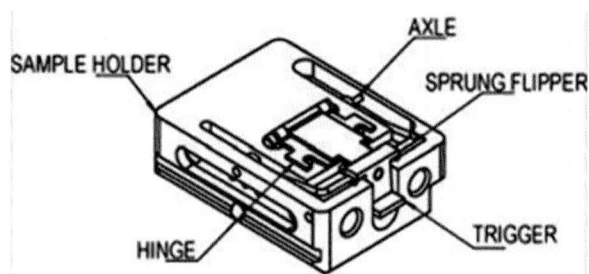

Fig. 1 Mousetrap design used for the freeze fracture system. The sample is trapped between two metal plates connected by a hinge, which is sprung open when triggered. Reproduced from [5] with permission from John Wiley and Sons

2. Aspirate culture media.

3. Wash substrate in 0.15 M ammonium formate for 1 min (*see* **Note 4**).

4. Sandwich cells between two metal plates connected by a hinge and rapidly freeze in liquid nitrogen-cooled propane (*see* Fig. 1).

5. Transfer sample to a liquid nitrogen flask containing the mousetrap device where the sample is mechanically fixed in place.

6. Transfer the sample to the instrument through the use of a glove box purged with argon gas to prevent frosting.

7. Sample is mounted directly onto a precooled sample insertion stage and transferred into the preparatory chamber.

8. Sample is fractured in the instrument at 168 K to minimize deposition of water. The trap is sprung by a transfer arm revealing cryogenically preserved cells (*see* **Note 5**).

9. Sample is transferred to a cold stage in the analysis chamber and held below 150 K throughout analysis (*see* **Note 6**).

3.3 Frozen Hydrate [13]

1. To prepare cells for analysis, add the desired amount of cells to a petri dish containing clean and dry silicon shards and allow growth in incubator until optimal coverage is obtained, usually about 24 h (*see* **Note 7**).

2. Warm 0.15 M ammonium formate solution (pH ~7.3) to 37 °C.

3. Cool the sample holder in liquid nitrogen after blowing it dry with nitrogen gas. Allow the holder to cool in the liquid nitrogen and do not add the sample until the boiling has stopped. Keep the holder submerged in liquid nitrogen for the duration of the sample preparation (*see* **Note 8**).

4. Wash the cells three times for ~5 s, in three beakers of the 0.15 M ammonium formate solution, to remove residual salts present from the cell media for a total of nine washes (*see* **Note 4**).

5. Dry the shards with a very gentle stream of nitrogen. If the nitrogen pressure is too great, streaking of the cells will occur. Also, be sure to dry the tweezers prior to freezing the sample (*see* **Note 9**).

6. Quickly plunge freeze the shard with cell growth into liquid propane for about 3–4 s before transferring to the liquid nitrogen-covered sample holder. If transfer times are too slow, the sample will warm above ~165 K and the cells will rupture.

7. Transfer into instrument for analysis, again, minimizing the amount of time that the cells are exposed to room temperature. The instrument should be operated with a precooled stage.

3.4 Chemical Fixation with Glutaraldehyde [6]

1. Seed the hTERT-BJ1 onto the clean silicon shards and allow growth in an incubator to occur for up to 2 days.

2. Wash the cells with phosphate-buffered saline (PBS).

3. Fix the cells with 2.5 % glutaraldehyde for 15 min in 37 °C.

4. Wash away excess GA with PBS.

5. Gently dry with nitrogen.

6. Plunge freeze in liquid propane and isopentane (3:1). This allows for a high cooling rate, reducing water crystallization.

7. Allow to freeze-dry overnight at –80 °C and 10^{-6} mbar.

8. Allow to warm by 10 °C/h to 30 °C.

9. If freeze-drying did not occur in the SIMS instrument, transfer sample into the instrument and complete analysis.

3.5 Chemical Fixation with Paraformalin/ Formaldehyde [7]

1. To prepare cells for analysis, add the desired amount of cells to a 24-well plate containing clean and dry poly-L-lysine-coated silicon shards and allow growth in incubator for 24 h until optimal coverage is obtained (*see* **Note 3**).

2. Aspirate the media.

3. Wash with phosphate-buffered saline (PBS) three times.

4. Incubate the cells with 4 % formalin for 15 min at 4 °C.

5. Wash with PBS five times and then wash with water three times.

6. Return to PBS for storage for several hours.

7. Wash with 0.15 M ammonium formate for 1 min and allow to dry.

8. Transfer into the SIMS instrument and begin analysis.

3.6 Chemical Fixation with Trehalose [16]

1. Cells were allowed to grow for several days prior to fixation.

2. Incubate cells in 50 mM trehalose for several hours.

3. Rinse for ~5 s with PBS buffer containing 50 mM trehalose and 10–15 % glycerol by weight. The addition of glycerol ensures stronger adhesion of the cells to the substrate.

4. Place a 200-mesh grid, followed by a thin substrate on top of the sample, creating a sandwich, and freeze in liquid nitrogen. Since trehalose acts similarly to a cryopreservant, flash freezing is not necessary in this protocol.

5. Place samples under vacuum and leave at 10^{-2}–10^{-7} mbar overnight (at least 15 h) to allow freeze-drying to occur.

6. Fracture the sample by removing the upper substrate and quickly load into the instrument for analysis. Long delays after fracturing can result in ambient hydration of the sample.

3.7 Freeze-Dry [17]

1. To prepare cells for analysis, add the desired amount of cells to a petri dish containing clean and dry shards and allow growth in incubator until optimal coverage is obtained.

2. Remove cells grown on the desired substrate from the incubator.

3. Wash the cells three times each in three beakers of the ammonium formate solution to remove residual salts from the cell media (*see* **Note 4**).

4. Allow the shards to dry slightly in air.

5. Plunge freeze the shard in liquid propane.

6. Store under liquid nitrogen until freeze-drying procedure begins.

7. Place cryofixed sample into a vacuum chamber with pressures of 10^{-3}–10^{-6} mbar overnight or 24 h to remove water by sublimation. This is done at room temperature; however, the water should sublime before any melting can occur.

8. Transfer into ToF-SIMS for analysis.

3.8 Collecting Images

1. In all sample preparation methods using cryofixation (frozen hydrate and freeze fracture), ToF-SIMS analysis must be performed using a cold stage (<150 K). This ensures that frozen samples do not warm prior to their analysis, preventing the dissipation of volatile compounds and the rearrangement or disruption of chemicals and organelles in the cells. A cold stage also helps to minimize water deposition onto the sample.

2. In order to ensure that an unaltered surface is being analyzed, it is important to operate the SIMS experiment under the static limit (1×10^{13} ions/cm^2). The factors contributing to the ion dose are spot size, primary ion current, and pixel size. Each will be discussed below.

   (a) Spot size is the diameter of the primary ion beam at its focal point and is what limits the spatial resolution in SIMS analysis. It can be changed through adjusting the beam alignment, but it is important to note that smaller spot sizes generally result in lower primary ion currents.

(b) A high primary ion current allows for more rapid image collection, as the static limit is reached more quickly. However, in order for a high primary ion current to be possible, a larger spot size is created, thus decreasing the spatial resolution.

(c) The pixel size can be determined by dividing the field of view by the amount of pixels in the acquired image. A smaller number of pixels results in fast acquisition times, but limits the spatial resolution.

3. Mass range is also an important factor in SIMS analysis. The mass range over which a SIMS analysis is acquired is dependent on the goals of the researcher, with a standard range beginning at 10 amu and going to 1,000 amu.

4. As lateral resolution is often a problem in cellular SIMS analysis, it is helpful to match the pixel size to the spot size. After the spectrum has been obtained, it is possible to combine adjacent pixels and their mass spectra to increase signal intensity.

3.9 Data Analysis

1. One of the many factors that makes SIMS so suited to single-cell analysis is its ability to create a chemical map of compounds of interest, thus showing the distributions of specific chemicals. It is possible to produce an image of a specific component by selecting the mass of interest and plotting its intensity against the spatial position. The intensity is, generally, shown using a false-color scale.

2. The success of sample preparation methods in maintaining the integrity of the cells can be verified using the localization of certain compounds and their fragments known to exist within a cell. Compounds including phosphocholine head group (PC, m/z 184), cholesterol (m/z 369), and sphingomyelin (SM) should only be found in the cellular region. The distribution of sodium and potassium ions also provides a good indication of the integrity of the cell. Sodium should only be found outside of the area of the cell, while potassium should be found in the cell regions [18]. If co-localization of sodium and potassium is seen, this is evidence for a ruptured cell membrane that may have resulted from inadequate sample preparation.

3. Adequate signal intensity is crucial for creating usable chemical maps. Signal intensity depends on a number of factors including the concentration of compound on the sample surface, the topography of the sample, the matrix surrounding the sample, and the stability of the primary ion beam. For these reasons, quantitative analysis is difficult.

4   Notes

1. It is crucial to work with only clean shards, as any contaminants can compromise SIMS results.

2. Handle shards with acetone-rinsed tweezers to minimize any contamination risk.

3. Poly-L-lysine improves the adherence of cells to the substrate surface with cells that do not normally adhere to solid surfaces.

4. The residual salts from the cell culture media obscure the SIMS signal from components of interest. Washing with a hypo-osmotic solution such as water will cause cells to rupture and so an iso-osmotic wash is necessary. Washing with ammonium formate or ammonium acetate has been found to remove these interfering salts without damaging the cell. These solutions also leave very little or no interfering residue on the cells after washing [4].

5. The fracture temperature is important in obtaining a good sample analysis. If the fracture temperature is too low, the ice signal will dominate the spectra, as condensation dominates and layers of ice accumulate on the surface. If the fracture temperature is too high, no ice will be present; however, chemical compounds, particularly lipids, will appear smeared, because sublimation dominates in this process.

6. As the sample has already been precooled in LN_2 prior to insertion in the instrument and precautions have been taken to prevent exposure to a damp atmosphere, cooling times will be reduced and sample degradation minimalized.

7. Analysis has been shown to be more successful when only 30–40 % of the silicon shard is covered with cells [13].

8. This sample block preparation minimizes the water deposition on the surface of the sample block after cooling, and thus on the sample. Ensure that the sample block is completely covered with LN_2 to prevent collection of ice on all exposed parts.

9. Having dry tweezers prior to plunge-freezing minimizes the amount of water that may deposit on the sample surface.

Acknowledgments

This project was supported by grants from the National Center for Research Resources (5P41RR031461) and the National Institute of General Medical Sciences (8 P41 GM103391) from the National Institutes of Health. In addition, infrastructure support from the National Science Foundation under grant number CHE-0908226 and by the Division of Chemical Sciences at the Department of Energy grant number DE-FG02-06ER15803 is acknowledged.

References

1. Roddy TP, Cannon DM, Meserole CA, Winograd N, Ewing AG (2002) Imaging of freeze-fractured cells with in situ fluorescence and time-of-flight secondary ion mass spectrometry. Anal Chem 74(16):4011–4019

2. Piehowski PD, Kurczy ME, Willingham D, Parry S, Heien ML, Winograd N, Ewing AG (2008) Freeze-etching and vapor matrix deposition for ToF-SIMS imaging of single cells. Langmuir 24(15):7906–7911

3. Vickerman JC (2011) Molecular imaging and depth profiling by mass spectrometry-SIMS, MALDI or DESI? Analyst 136(11):2199–2217

4. Berman ESF, Fortson SL, Checchi KD, Wu L, Felton JS, Wu KJJ, Kulp KS (2008) Preparation of single cells for imaging/profiling mass spectrometry. J Am Soc Mass Spectrom 19(8):1230–1236

5. Fletcher JS, Rabbani S, Henderson A, Lockyer NP, Vickerman JC (2011) Three-dimensional mass spectral imaging of HeLa-M cells: sample preparation, data interpretation and visualisation. Rapid Commun Mass Spectrom 25(7):925–932

6. Malm J, Giannaras D, Riehle MO, Gadegaard N, Sjovall P (2009) Fixation and drying protocols for the preparation of cell samples for time-of-flight secondary Ion mass spectrometry analysis. Anal Chem 81(17):7197–7205

7. Rabbani S, Fletcher JS, Lockyer NP, Vickerman JC (2011) Exploring subcellular imaging on the buncher-ToF J105 3D chemical imager. Surf Interface Anal 43(1–2):380–384

8. Parry S, Winograd N (2005) High-resolution TOF-SIMS imaging of eukaryotic cells preserved in a trehalose matrix. Anal Chem 77(24):7950–7957

9. Hoppert M, Holzenburg A (1998) Electron microscopy in microbiology. BIOS Scientific Publishers, Oxford

10. Severs NJ, Newman TM, Shotton DM (1995) A practical introduction to rapid freezing techniques. In: Severs NJ, Shotton DM (eds) Rapid freezing, freeze fracture and deep etching. Wiley-Liss, New York, NY

11. Lanekoff I, Kurczy ME, Adams KL, Malm J, Karlsson R, Sjovall P, Ewing AG (2011) An in situ fracture device to image lipids in single cells using ToF-SIMS. Surf Interface Anal 43(1–2):257–260

12. Brison J, Benoit DSW, Muramoto S, Robinson M, Stayton PS, Castner DG (2011) ToF-SIMS imaging and depth profiling of HeLa cells treated with bromodeoxyuridine. Surf Interface Anal 43(1–2):354–357

13. Piwowar AM, Keskin S, Delgado MO, Shen K, Hue JJ, Lanekoff I, Ewing AG, Winograd N (2013) C60-ToF SIMS imaging of frozen hydrated HeLa cells. Surf Interface Anal 45(1):302–304

14. Piwowar AM, Fletcher JS, Kordys J, Lockyer NP, Winograd N, Vickerman JC (2010) Effects of cryogenic sample analysis on molecular depth profiles with TOF-secondary ion mass spectrometry. Anal Chem 82(19):8291–8299

15. Roddy TP, Cannon DM, Ostrowski SG, Ewing AG, Winograd N (2003) Proton transfer in time-of-flight secondary ion mass spectrometry studies of frozen-hydrated dipalmitoylphosphatidylcholine. Anal Chem 75(16):4087–4094

16. Braun RM, Blenkinsopp P, Mullock SJ, Corlett C, Willey KF, Vickerman JC, Winograd N (1998) Performance characteristics of a chemical imaging time-of-flight mass spectrometer. Rapid Commun Mass Spectrom 12(18):1246–1252

17. Fletcher JS, Lockyer NP, Vaidyanathan S, Vickerman JC (2007) TOF-SIMS 3D biomolecular imaging of Xenopus laevis oocytes using buckminsterfullerene (C-60) primary ions. Anal Chem 79(6):2199–2206

18. Stryer L (1981) Biochemistry, 2nd edn. W. H. Freeman and Company, San Francisco, CA

Chapter 3

TOF-SIMS Imaging of Lipids on Rat Brain Sections

David Touboul and Alain Brunelle

Abstract

Since several decades, secondary ion mass spectrometry (SIMS) coupled to time of flight (TOF) is used for atomic or small inorganic/organic fragments imaging on different materials. With the advent of polyatomic ion sources leading to a significant increase of sensitivity in combination with a reasonable spatial resolution (1–10 μm), TOF-SIMS is becoming a more and more popular analytical platform for MS imaging. Even if this technique is limited to small molecules (typically below 1,000 Da), it offers enough sensitivity to detect and locate various classes of lipids directly on the surface of tissue sections. This chapter is thus dedicated to the TOF-SIMS analysis of lipids in positive and negative ion modes on rat brain tissue sections using a bismuth cluster ion source.

Key words Secondary ion mass spectrometry, Time-of-fight, Mass spectrometry imaging, Lipid, Rat brain

1 Introduction

Secondary ion mass spectrometry (SIMS) coupled with magnetic double-focusing sector field mass spectrometer was firstly used by Castaing and Slodzian in 1962 for microanalysis and mass spectrometry imaging (MSI) [1]. Until the beginning of the twenty-first century, SIMS imaging was limited to atom or small fragment analysis by the use of monoatomic primary ion sources (In^+ or Ga^+) even if high lateral resolution (until 10 nm) was achievable [2, 3]. Thanks to fundamental works in physics in the 1990s [4, 5], polyatomic primary ion sources have become commercially available. The first generation was based on gold clusters (Au_3^+) allowing the detection of intact lipid species on rat brain sections in the positive and negative ion modes [6, 7]. At the same time, a second generation of cluster ion sources was developed. The first one was based on the vaporization of fullerene C_{60} powder and formation of singly and doubly charged C_{60} which are then focalized through an ion optic column at the sample surface. Even if the C_{60} source offers a very low surface damage and can be efficiently

Lin He (ed.), *Mass Spectrometry Imaging of Small Molecules*, Methods in Molecular Biology, vol. 1203,
DOI 10.1007/978-1-4939-1357-2_3, © Springer Science+Business Media New York 2015

used for depth profiling, the best spatial resolution is restricted to 10 μm in routine analysis [8]. At the contrary, bismuth cluster (Bi_3^+ and Bi_3^{2+}) liquid metal ion gun (LMIG) offers a good compromise between sensitivity and lateral resolution (below 2 μm in routine analysis using the high-current bunched mode and down to 400 nm using the burst alignment mode) [9]. The following protocol describes how to prepare a rat brain section for TOF-SIMS imaging of lipids using a bismuth cluster ion source. A more detailed description of cluster TOF-SIMS imaging can be found in a tutorial paper [10].

2   Materials

2.1   Chemicals

1. Sodium pentobarbital.
2. Optimal cutting temperature (OCT) embedding medium.
3. Dry ice.

2.2   Instruments and Materials

1. TOF-SIMS IV (ION-TOF GmbH, Münster, Germany) or equivalent. The data acquisition and processing software is SurfaceLab 6.3 (ION-TOF GmbH, Münster, Germany) or later.
2. Cryostat (model CM3050-S; Leica Microsystems SA, Nanterre, France, or equivalent).
3. Optical microscope (Olympus BX 51 fitted with ×1.25 to ×50 lenses) (Olympus France SAS, Rungis, France) equipped with a Color View I camera, monitored by Cell[B] software (Soft Imaging System GmbH, Münster, Germany) or equivalent.
4. Desiccator.
5. Sample plate for TOF-SIMS imaging (conductive glass slide coated with indium-tin-oxide, stainless steel plate, or silicon wafer).

3   Methods

3.1   Tissue Sectioning

All experiments on rat brains were performed in accordance with the protocols approved by the National Commission on animal experimentation and by the recommendations of the European commission DGXI.

1. Euthanize male Wistar rats of typical 400 g weight by an intraperitoneal injection of sodium pentobarbital (>65 mg/kg).
2. Freeze the trimmed tissue blocks immediately in dry ice to prevent crack formation during freezing and store at –80 °C prior to MS experiments (*see* **Note 1**).

3. Add a few drops of OCT embedding medium and quickly deposit the tissue block over it to fix the frozen tissue block to the cryostat adapter. The transparent OCT embedding medium slowly turns white when hardens.

4. Cut tissue section of 8–15 μm thickness using the cryostat at −20 °C.

5. Deposit the tissue sections on either a conductive glass slide, a stainless plate, or a silicon wafer adapted to the TOF-SIMS holder (*see* **Note 2**).

6. Store the samples at −80 °C in a box to prevent degradation by ice.

7. Warm the sample at room temperature for a few seconds before MS analysis, and dry the sample under vacuum at a pressure of a few hPa for 30 min in a desiccator.

8. Take pictures of the sample using a microscope (×1.5, ×10, and ×20). It is important to keep the same orientation of the sample between the microscope observation and the TOF-SIMS acquisition in order to easily correlate the images.

3.2 Ion Source Optimization for High-Current Bunched Mode

1. Introduce the sample holder in the mass spectrometer. Before starting the primary ion gun, the vacuum in the analysis chamber must be below 2.10^{-6} hPa (The vacuum is read directly on the vacuum control panel).

2. Open the Navigator, Analyzer, and LMIG panel. The jpg image recorded by the microscope (or by a scanner) can be co-registrated in the Navigator. The user can then navigate directly on the section and easily find small histological structures.

3. Open the Spectra and Image panel.

4. Switch on the high voltage required for the analyzer and LMIG.

5. Start the LMIG using the batch command provided by the manufacturer. It allows an automatic switch-on of the LMIG source. The operation takes about 10 min.

6. Measure the primary ion current using the Faraday cup provided on each sample holder. First of all, the primary ion beam is automatically centered in order to optimize the ion transmission through the ion optics. The ion current is then measured using the direct current (DC) mode. In this mode, the primary ions are neither bunched nor pulsed and all the primary ions (Bi^+, Bi_3^+, Bi_3^{2+}, Bi_5^+, Bi_5^{2+}, Bi_7^+ …) are focused on the surface. Typical DC current between 12 and 15 nA is measured. The values of the heating current (about 2.85 A), emission current (about 1 μA), suppressor voltage (about 1,000 V), extractor voltage (9,000 V), and lens source (about 3,000 V) are always reported in the laboratory book in order to check any voltage deviation related to any source problem. The pulsed primary

ion current is then measured for Bi$^+$ (about 1.4 pA) and Bi$_3^+$ (about 0.4 pA) using the pulsed mode and a cycle time analysis of 100 μs. These values are necessary when calculating the primary ion dose density (also called "fluence") before an MSI experiment.

7. Switch on the electron flood gun used for insulating samples.

8. In the positive ion mode, the primary ion beam is aligned with the optical images using an "A-Grid," i.e., a metallic grid where the A-letter is drawn. For that purpose, X- and Y-Target parameters need to be tuned in order to get a perfect match between the optical and ionic images.

9. Move the stage to the sample, and adjust the Z-value, *i.e.*, the distance between the sample and the extraction cone, in order to get the best sensitivity.

10. Adjust the Charge Compensation parameter directly on the sample surface. The Z value needs to be checked and can vary from the "A-grid." For insulating samples, such as a brain rat section, the optimization of this parameter is highly important to avoid significant reduction in the mass resolution. If needed, X- and Y-Target parameters can be optimized on the tissue sample to reach a perfect match.

11. Redo **steps 8–10** in the negative ion mode.

12. Spatial resolution can be checked by acquiring a profile on the "A-grid." A typical value is between 2 and 3 μm.

13. Save all parameters before starting an imaging experiment.

3.3 TOF-SIMS Imaging of Lipids on a Rat Brain Section

1. Before any imaging experiments, a mass spectrum needs to be acquired on the sample surface with a very low dose to avoid damage to the sample surface.

2. Calibrate the mass spectrum. In the positive ion mode, the mass calibration is initially made with H^+, H_2^+, H_3^+, C^+, CH^+, CH_2^+, CH_3^+, and $C_2H_5^+$ ions. In the negative ion mode, the mass calibration is initially made with H^-, C^-, CH^-, CH_2^-, CH_3^-, C_2^-, C_3^-, and C_4H^- ions. To further improve mass accuracy, the mass calibration can be refined by adding ions of higher mass, such as fatty acid carboxylate ions and deprotonated vitamin E in the negative ion mode. Mass resolution of about 5,000 at m/z 500 can be achieved on the tissue section with mass accuracy of about 20 ppm over the complete mass range (m/z 0–1,000).

3. Some peaks of reference can be selected in order to reconstruct the ion image during the acquisition and thus allow structural characterization of the tissue section. A rather complete list of lipid species already observed by TOF-SIMS on tissue section is

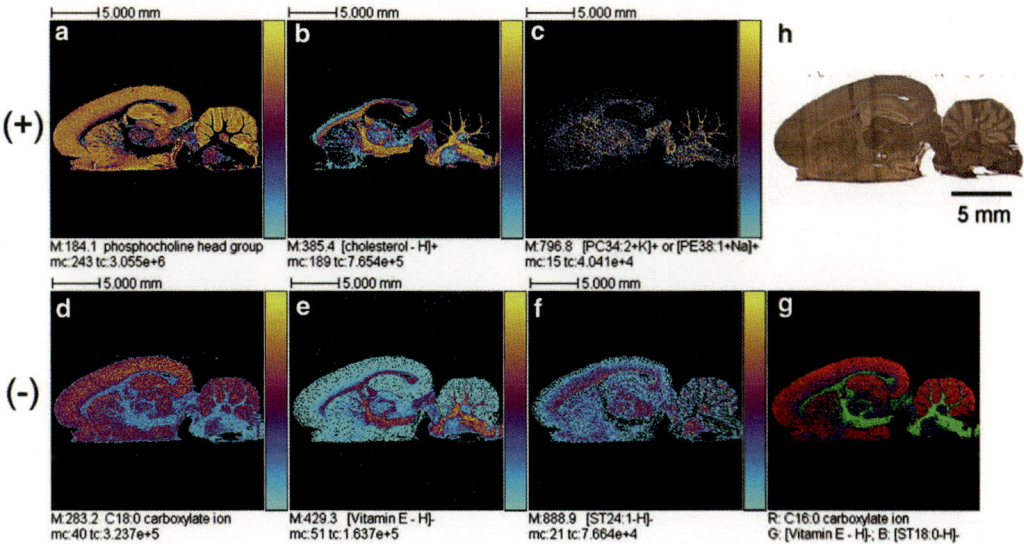

Fig. 1 TOF-SIMS ion images of a complete sagittal rat brain section (22.4×22.4 mm^2). (**a**–**c**) TOF-SIMS ion images in the positive ion mode. (**a**) m/z 184.1 (phosphocholine head group); (**b**) m/z 385.4 ([cholesterol-H]$^+$); (**c**) m/z 796.8 ([PE 40:4 + H]$^+$, [PE38:1 + Na]$^+$, or [PC34:2 + K]$^+$). (**d**–**g**) TOF-SIMS ion images in negative ion mode. (**d**) m/z 283.2 (fatty acid carboxylate [C18:0 – H]$^-$); (**e**) m/z 429.3 ([vitamin E–H]$^-$); (**f**) m/z 888.9 ([ST d18:1/24:1 – H]$^-$); (**g**) three-color overlay between TOF-SIMS ion images C16:0 carboxylate ion (*red*), vitamin E (*green*), and [ST d18:1/18:0 – H]$^-$ (*blue*). For all TOF-SIMS images: primary ion Bi$_3^{2+}$, 50 keV, 3.4×10^8 ions/cm^2, area 22.4×22.4 mm^2, 256×256 pixels, pixel size 87.5×87.5 µm^2. The amplitude of the color scale corresponds to the maximum number of counts, *mc*, and could be read as [0, mc]. *tc* is the total number of counts recorded for the specified m/z (it is the sum of counts in all the pixels). (**h**) An optical ion image of the sagittal rat brain section. Adapted with permission from ref. [12]

available [11]. Reference mass spectra of pure lipid compounds can also be found in the spectra library of SurfaceLab software.

4. Select the area of interest. When analyzing an area smaller than 500×500 µm^2, the sample is fixed and the primary ion beam scans the whole surface. For larger surfaces, the selection of a patch of 500×500 µm^2 (or smaller) image is required. Before each acquisition, the primary ion dose is calculated in order to keep it below the limit of static SIMS (1×10^{13} ions/cm^2).

5. Typical TOF-SIMS imaging of lipids from a complete sagittal rat brain section is provided in Fig. 1, as well as a close-up TOF-SIMS analysis of a 500 µm × 500 µm area in Fig. 2 [12].

4 Notes

1. Formaldehyde fixation is always avoided when possible in order to prevent chemical modification of the lipid species [13].

2. Classical histochemistry (hematoxylin and eosin staining for example) is usually performed on adjacent tissue sections in

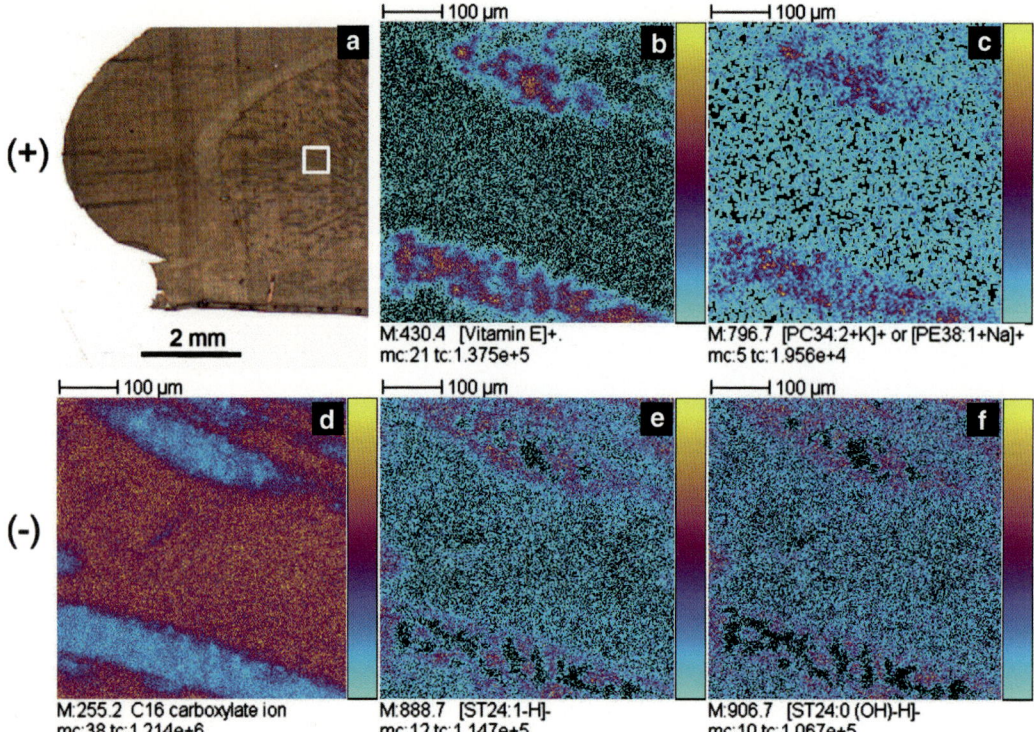

Fig. 2 TOF-SIMS ion images of a zoom-in sagittal rat brain section (500 × 500 μm²). (**a**) An optical picture of the sagittal rat brain section. Ion images were recorded in the square delimited in *white*. (**b**, **c**) TOF-SIMS ion images in the positive ion mode. (**b**) m/z 430.4 ([vitamin E]⁺). (**c**) m/z 796.7 ([PE 40:4 + H]⁺, [PE38:1 + Na]⁺, or [PC34:2 + K]⁺). (**d–f**) TOF-SIMS ion images in the negative ion mode. (**d**) m/z 255.2 (C16:0 carboxylate ion). (**e**) m/z 888.7 ([ST d18:1/24:1-H]⁻). (**f**) m/z 906.7 ([ST d18:1/24:0-OH-H]⁻). Primary ion Bi_3^{2+}, 50 keV, 2.5×10^{11} ions/cm², area 500 × 500 μm², 256 × 256 pixels, pixel size 2 × 2 μm². The amplitude of the color scale corresponds to the maximum number of counts, *mc*, and could be read as [0, mc]. *tc* is the total number of counts recorded for the specified m/z (it is the sum of counts in all the pixels). Adapted with permission from ref. [12]

order to visualize the different anatomical parts of the rat brain. It must be noticed that histochemistry can also be performed on the same tissue section that was used for TOF-SIMS experiments [14].

References

1. Castaing R, Slodzian GJ (1962) Optique corpusculaire-premiers essais de microanalyse par émission ionique secondaire. J Microsc 1:395–399

2. Hallegot P, Girod C, Levi-Setti R (1990) Scanning ion microprobe assessment of biological sample preparation techniques. Scanning Microsc 4(3):605–612

3. Pacholski ML, Winograd N (1999) Imaging with mass spectrometry. Chem Rev 99(10):2977–3006

4. Benguerba M, Brunelle A, Della-Negra S et al (1991) Impact of slow gold clusters on various solids: nonlinear effects in secondary ion emission. Nucl Instr Meth Phys Res B 62(1):8–22

5. Brunelle A, Della-Negra S, Depauw J et al (2001) Enhanced secondary-ion emission under gold-cluster bombardment with energies from keV to MeV per atom. Phys Rev A 63(2):022902 1–10

6. Touboul D, Halgand F, Brunelle A et al (2004) Tissue molecular ion imaging by gold cluster ion bombardment. Anal Chem 76: 1550–1559

7. Sjövall P, Lausmaa J, Johansson B (2004) Mass spectrometric imaging of lipids in brain tissue. Anal Chem 76(15):4271–4278

8. Weibel D, Wong S, Lockyer N et al (2003) A C60 primary ion beam system for time of flight secondary ion mass spectrometry: its development and secondary ion yield characteristics. Anal Chem 75(7):1754–1764

9. Touboul D, Kollmer F, Niehuis E et al (2005) Improvement of biological time-of-flight-secondary ion mass spectrometry imaging with a bismuth cluster ion source. J Am Soc Mass Spectrom 16(10):1608–1618

10. Brunelle A, Touboul D, Laprévote O (2005) Biological tissue imaging with time-of-flight secondary ion mass spectrometry and cluster ion sources. J Mass Spectrom 40(8): 985–999

11. Passarelli M, Winograd N et al (2011) Lipid imaging with time-of-flight secondary ion mass spectrometry (ToF-SIMS). Biochim Biophys Acta 1811(11):976–990

12. Benabdellah F, Seyer A, Quinton L et al (2010) Mass spectrometry imaging of rat brain sections: nanomolar sensitivity with MALDI versus nanometer resolution by TOF-SIMS. Anal Bioanal Chem 396:151–162

13. Eltoum I, Fredenburgh J, Myers RB et al (2001) Introduction to the theory and practice of fixation of tissues. J Histotechnol 3(18):173–190

14. Bich C, Vianello S, Guérineau V (2013) Compatibility between TOF-SIMS lipid imaging and histological staining on a rat brain section. Surf Interface Anal 45:260–263

Chapter 4

MALDI-MS-Assisted Molecular Imaging of Metabolites in Legume Plants

Erin Gemperline and Lingjun Li

Abstract

Mass spectrometric imaging (MSI) is a powerful analytical tool that provides spatial information of several compounds in a single experiment. This technique has been used extensively to study proteins, peptides, and lipids, and is becoming more common for studying small molecules such as endogenous metabolites. With matrix-assisted laser desorption/ionization (MALDI)-MSI, spatial distributions of multiple metabolites can be simultaneously detected within a biological tissue section. Herein, we present a method developed specifically for imaging metabolites in legume plant roots and root nodules which can be adapted for studying metabolites in other legume organs and even other biological tissue samples. We focus on essential steps such as sample preparation and matrix application, comparing several useful techniques, and present a standard workflow that can be easily modified for different tissue types and instrumentation.

Key words Imaging mass spectrometry, Mass spectrometric imaging, MALDI, TOF/TOF, Legume, Metabolite, Mass spectrometry

1 Introduction

Legumes are extremely important to the agriculture industry because they can grow in a wide variety of agroecological conditions and have developed the unique ability to fix their own nitrogen through their symbiotic relationship with soil bacteria known as rhizobia [1]. Decades of research on legumes have been dedicated to deciphering the metabolic networks involved in nitrogen fixation [2, 3]. A major technical challenge when studying biological systems is to study metabolomic pathways without affecting them [4]. Most of the techniques currently in use for studying plant metabolomics rely on plant extracts which destroy the tissue samples and thus eliminate the ability to determine analyte distribution within the tissue. Spatially imaging the metabolome would allow major progress in the understanding of the coordination between rhizobia and legumes, and can be applied toward unraveling other metabolic pathways within the plant.

Lin He (ed.), *Mass Spectrometry Imaging of Small Molecules*, Methods in Molecular Biology, vol. 1203, DOI 10.1007/978-1-4939-1357-2_4, © Springer Science+Business Media New York 2015

MALDI-MSI allows for direct analysis of intact tissues that enables sensitive detection of analytes in single organs [5, 6] and even single cells [7, 8]. MALDI-MSI eliminates the steps of sample extraction, purification, and separation which are time-consuming steps in traditional MS experiments [9, 10]. Another feature of MALDI-MSI is its ability to image a broad mass range of molecules, from small molecules to large proteins, without requiring prior knowledge of the target analytes. Tandem mass (MS/MS) experiments can be performed directly on the tissue with MALDI-MSI to elucidate the identities of unknown metabolites that may play important roles in biological mechanisms and pathways.

Herein we present a simple scheme to map metabolites in legume roots and root nodules that can be adapted and applied to various tissue types. After dissection, the root nodule is sliced and mounted on an indium tin oxide (ITO)-coated glass slide. The tissue is then desiccated and covered with matrix via either solution-based spraying using airbrush or automated matrix sprayer or dry matrix application through sublimation. An array of mass spectra is collected by rastering across the surface of the tissue section via moving the MALDI sample stage in predefined x–y coordinates against a fixed laser beam irradiation. Processing this array of spectra produces a cohesive MS image wherein each pixel contains the information of a corresponding spectrum. This workflow is outlined in Fig. 1.

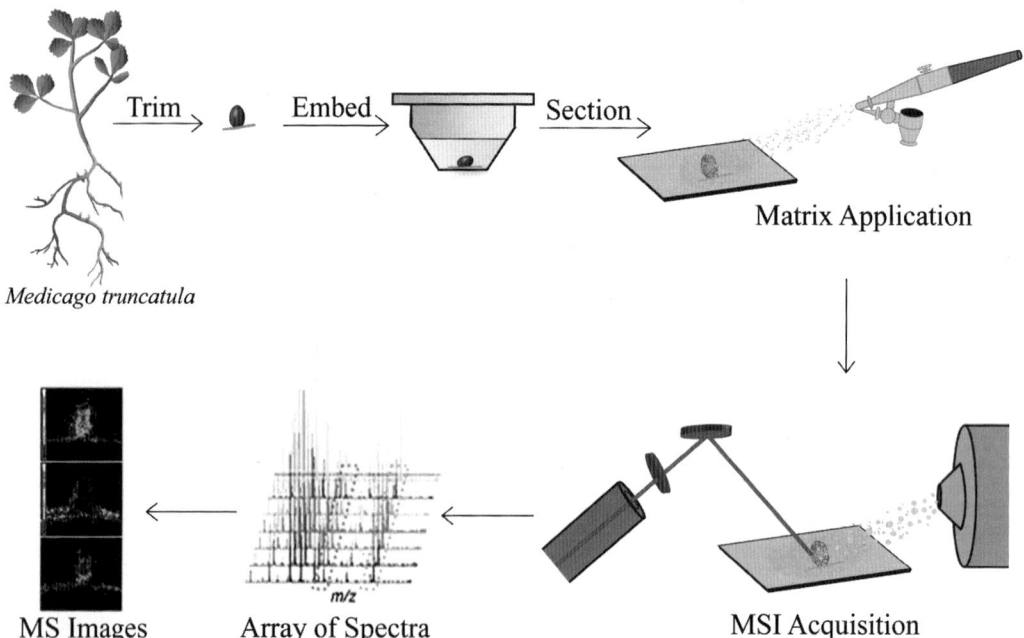

Fig. 1 General scheme of MSI technology and its application to mapping plant metabolites. Acquisition occurs by collecting mass spectra for each pixel and processing this array of spectra into representative 2D images of specific *m/z* values

2 Materials

2.1 Reagents/ Equipment

1. Gelatin (100 mg/mL in deionized water).

2. Cryostat.

3. 25–75–0.8 mm (width-length-thickness) indium tin oxide (ITO)-coated glass slides.

4. 2,5-Dihydroxybenzoic acid (DHB): 150 mg/mL in 50% methanol/0.1% formic acid v/v (airbrush) or 40 mg/mL in 50% methanol/0.1% formic acid v/v (automatic sprayer).

5. Deionized water.

6. Methanol.

7. Airbrush coupled with 75 mL steel container.

8. TM-Sprayer system (HTX Technologies, LLC, Carrboro, NC, USA) (*see* **Note 1**).

9. Sublimation apparatus.

10. Scanner.

2.2 Instrumentation

1. An ultrafleXtreme MALDI-TOF/TOF (Bruker Daltonics, Billerica, MA, USA) analyzer equipped with a 2 kHz FlatTop smartbeam-II™ Nd:YAG laser (spot diameter down to 10 μm) can be used for imaging. Other types of MALDI mass spectrometers may also be used. Acquisitions can be performed in positive or negative ion reflectron mode depending on the sample type.

2. Instrument parameters for a Bruker MALDI-TOF/TOF can be set using flexImaging and flexControl software (Bruker Daltonics). To produce ion images, spectra can be generated by averaging 500 laser shots over the mass range and collected at 25–100 μm intervals in both the x and y dimensions across the surface of the sample. The mass spectra can be externally calibrated using DHB matrix peaks or external standards applied directly to the glass slide.

3 Methods

3.1 Tissue Preparation

Sample preparation is a crucial step in producing reproducible and reliable mass spectral images. The quality of the images greatly depends upon factors such as tissue fixing methods, embedding medium, and slice thickness. Microwave irradiation and heat denaturation with the Denator Stabilizer T1 (Gothenburg, Sweden) have been reported to prevent postmortem protein degradation by deactivating proteolytic enzymes [11]. For imaging applications, ideal section thickness should be the width of one cell; 8–20 μm thickness is appropriate. Thicker sections tend to have better tissue integrity (i.e., less tearing or folding) but thinner sections typically result in better sensitivity.

1. Trim the root nodule from the plant, leaving 3–4 mm of root attached to the nodule (1–2 mm on each side of the nodule).

2. Immediately after dissection, submerge the tissue in a small plastic cup of gelatin (*see* **Notes 2** and **3**).

3. Use forceps to orient the tissue as desired.

4. Once the tissue is stuck to the bottom of the cup, use a syringe or pipet to cover the tissue with more gelatin (approximately 3–5 mm higher than the tissue) (*see* **Note 4**). Make sure that the tissue is completely surrounded by gelatin on all sides and there are no bubbles present in the gelatin (*see* **Note 5**).

5. Flash freeze the tissue by placing the cup in a dry ice/ethanol bath or gradually freeze by placing the cup on dry ice only until the gelatin hardens and becomes opaque (*see* **Note 6**).

6. Store tissue in −80 °C freezer until use.

7. Remove frozen tissue from the −80 °C freezer, cut away the plastic cryostat cup, and trim excess gelatin (approximately 3–4 mm on each side of the tissue). Mount the embedded tissue to the cryostat chuck with a dime-sized amount of optimal cutting temperature (OCT) media (*see* **Note 7**). Place in cryostat box until the OCT solidifies.

8. Prior to cryostat slicing, allow the chuck and gelatin to equilibrate in the cryostat box to the appropriate temperature (approximately 15 min) (*see* **Note 8**).

9. Cryostat section slices with 8–20 μm thickness as appropriate (*see* **Note 9**).

10. Thaw mount each slice onto the ITO-coated glass slide (or MALDI plate) (*see* **Notes 10** and **11**). If the tissue is too large for one glass slide (i.e., whole-body slices) you may position the section on multiple slides and digitally put the images together.

11. For 3D imaging, obtain multiple slices that are evenly distributed throughout the z-axis of the tissue and thaw mount each section individually onto the ITO-coated glass slide(s) (*see* **Note 12**).

12. Place the ITO-coated glass slides with the tissue sections in a desiccator for at least 30 min before matrix application.

3.2 Matrix Application

MALDI requires deposition of an organic, crystalline compound, typically a weak acid, on the tissue of interest to assist analyte ablation and ionization [12]. Choosing a MALDI matrix and its application method is essential for quality mass spectrometric imaging experiments. Conventional matrices include CHCA (α-cyano-4-hydroxycinnamic acid) and DHB. Less traditional matrices such as DAN (1,5-diaminonaphthalene), DMAN

(1,8-bis(dimethylamino)naphthalene), DHPT (2,3,4,5-tetra(3′,4′-dihydroxylphenyl)thiophene), TiO_2 nanoparticles, and ionic matrices are being used and are reported to improve spectral quality, crystallization, and vacuum stability [13–17]. Different matrices provide different amounts of coverage, signal intensity, matrix interference, and ionization efficiency. It is important to choose a matrix that gives the best results for the particular analytes of interest. The matrix application technique also plays a role in the quality of mass spectral images. Three matrix application methods are presented here: airbrush, automatic sprayer, and sublimation. Airbrush matrix application has been widely used in MALDI imaging and is relatively fast and easy, but is less reproducible and sometimes causes diffusion of analytes [18]. Automatic sprayer systems, like the TM-Sprayer, have been developed which remove the variability seen with manual airbrush application, making the spray more reproducible, but is more time consuming. Sublimation is a dry matrix application technique that is becoming more and more popular for mass spectral imaging of metabolites and small molecules [19]. Sublimation reduces analyte diffusion, but lacks the solvent necessary to observe higher mass compounds.

3.2.1 Airbrush Application of MALDI Matrix

1. Thoroughly clean the airbrush solution container and nozzle with methanol every time before matrix application.

2. Fill the solution container with DHB matrix solution (150 mg/mL in 50% methanol/0.1% formic acid v/v) and place the airbrush approximately 35 cm from the glass slide (*see* **Note 13**).

3. Apply 10–15 coats of matrix on the surface of the slide with a spray duration of 10 and 30 s drying time in between each coat.

4. Thoroughly clean the airbrush again with methanol when finished to avoid clogging from the matrix solution.

3.2.2 Automatic Sprayer Application of MALDI Matrix

1. Start compressed nitrogen-flow to the TM-Sprayer to 10 psi.

2. Turn on the TM-Sprayer, and set valve to the "Load" position.

3. Set the sprayer method with the TM-Sprayer software (*see* **Note 14**) and turn on the solvent pump to approximately 0.250 mL/min.

4. Use a syringe to inject your matrix solution (40 mg/mL DHB in 50% methanol/0.1% formic acid v/v) into the sample loop with 20% overfill.

5. Place ITO-coated glass slide with tissue slices into the spray chamber.

6. Switch valve to "Spray" and wait for 1–2 min for the matrix to reach the nozzle tip.

7. Start sprayer method.

8. When finished, switch valve back to "Load" keeping the pump flow on and flush the loop with 50 % methanol three times.

9. Turn the temperature back down to 30 °C. Wait for the temperature to come to at least 50 °C and turn off the pump, nitrogen, and sprayer system.

3.2.3 Sublimation
Application of MALDI
Matrix

1. Weigh out 300 mg DHB into the bottom of the sublimation chamber as shown in Fig. 2a.

2. Use double-sided conductive tape to stick the glass slide to the underside of the cold finger (top portion of the sublimation

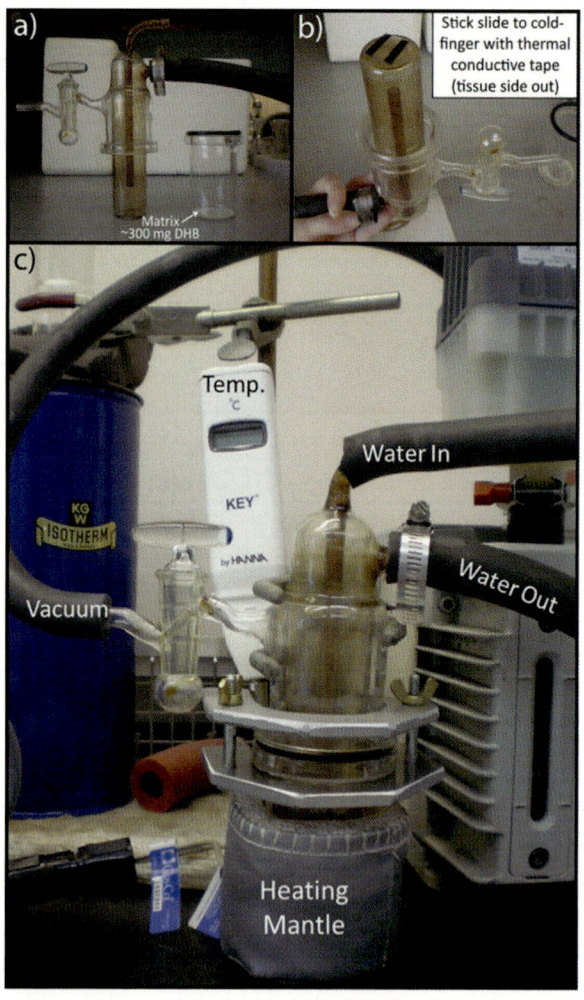

Fig. 2 Labeled photograph of sublimation apparatus setup. (**a**) Weigh out the matrix and place it in the *bottom portion* of the sublimation chamber. (**b**) The ITO slide with the tissue slice on it is attached to the underside of the condenser by thermal conducting tape. (**c**) Overall setup of the sublimation chamber connected to water, vacuum pump, and heating mantel and monitoring the temperature

chamber), with the tissue sections facing down (*see* **Note 15**) as shown in Fig. 2b.

3. Clamp the top and bottom of the sublimation chamber together and connect the vacuum and water as shown in Fig. 2c.

4. Place sublimation chamber in a heating mantle that is at room temperature.

5. Turn on the vacuum pump. After 15 min turn on the water. Wait for an additional 5 min and turn on the heating mantle.

6. The heating mantle should reach 120 °C over the course of 10 min.

7. After 10 min, turn off heat and water, close valve to vacuum (so the inside of the chamber remains under vacuum), and turn off the vacuum pump.

8. Allow the chamber to come to room temperature before releasing the vacuum pressure and removing the glass slide.

3.3 Image Acquisition

1. Mark a + pattern on each corner of the slide with a WiteOut correction fluid pen. Place the glass slide into the MALDI slide adapter plate and use a scanner to scan an optical image of the slide and samples (*see* **Note 16**).

2. Set up an image acquisition file using flexControl (Bruker Daltonics) (*see* **Note 17**).

3. Load the optical image into flexImaging and toggle back and forth between flexImaging and flexControl to set the three "teach points" which will align the plate with the optical image (*see* **Note 18**).

4. Calibrate the instrument with the "calibration" tab in flexControl. Common matrix peaks or external standards may be used for calibration (*see* **Note 19**).

5. Specify the areas you would like to image by tracing around the tissue slice with the "add polygon measurement region" tool. Also trace around a small spot of pure matrix for comparison (*see* **Note 20**).

6. Start automatic run. The software will calculate the estimated time for completing the experiment. This can be seen at the top of the "regions" window.

7. For 3D imaging, perform acquisition as described above for all tissue sections.

3.4 Image Processing

3.4.1 2D Image Generation

1. Open flexImaging software and open your imaging file. This could take several minutes depending on the processing power of your computer. The spectrum display will show the average spectrum of all collected spectra.

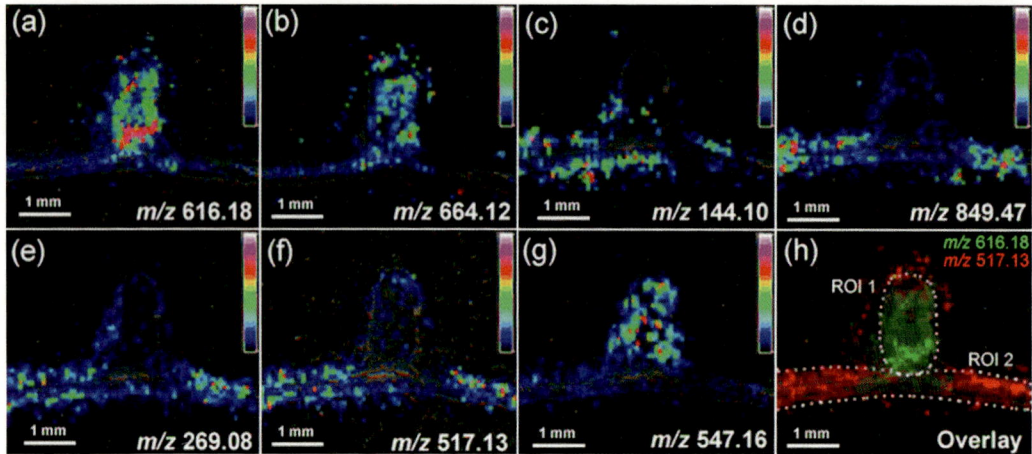

Fig. 3 Representative metabolite distribution in *Medicago truncatula* root nodule section revealed by MALDI-MSI. (**a**) Heme-moiety (*m/z* 616.15), (**b**) NAD (*m/z* 664.10), (**c**) proline betaine (*m/z* 144.10), (**d**) a putative sodiated lipid (*m/z* 849.47), (**e**) formononetin (*m/z* 269.08), (**f**) formononetin MalGlc (*m/z* 517.13), (**g**) afrormosin MalGlc (*m/z* 547.16) displaying distinct distribution patterns in roots and nodules. (**h**) An overlaid image of (**a**) and (**f**). Adapted with permission from ref. [20]

2. You can click on peaks on the mass spectrum with the "mass filter selection" tool to see where the analyte is distributed within the tissue section. The "color gradient" tool will let you see the intensity of the ion within that tissue section. You may also use the "show single spectrum" tool to click on a pixel on the imaging display and see the mass spectrum for that specific location on the tissue.

3. You may view the average spectra for several different regions at once by checking the "spectrum" boxes on the region window of the regions you would like to compare, right clicking on the spectrum display, and selecting display type 2D stack plot.

4. Images and spectra can be saved by using the edit copy function and pasting the image in Microsoft PowerPoint or similar programs. Figure 3 shows MSI detection of several metabolites displaying distinct distribution patterns in the root and nodule of *Medicago truncatula*, a model legume plant [20].

5. MSiReader is an open-source software available on www. MSiReader.com and can be used for automatic image generation and more automated data processing [21].

3.4.2 3D Image Generation

1. Open flexImaging and highlight your peak of interest. Save images of serial sections as picture files (i.e., JPEG).

2. Open Image J software (http://rsbweb.nih.gov/ij/, NIH) (*see* **Note 21**).

3. Open the consecutive series of tissue images that you saved with flexImaging (*see* **Note 22**).

4. Align the orientation and position of each image and make slight adjustments to get the images aligned using the "translate" and "rotate" functions under "image."

5. Under "image," "stacks," open "images to stack" to combine the 2D images into one stack.

6. To view the stack in three dimensions, open "image," "stacks," and click on "3D project." Parameters such as axis of rotation and slice spacing can be adjusted by changing values in the 3D projection window.

4   Notes

1. The TM-Sprayer from HTX Technologies is a highly specialized automatic matrix sprayer. Other matrix sprayers are available; however they have different features than the TM-Sprayer and should be used according to the manufacturer's recommendations.

2. Gelatin is used for embedding the tissues rather than OCT media because OCT contains a high concentration of polyethylene glycol (PEG) which produces interfering peaks in the mass spectral analysis.

3. Before placing the tissue in the cup, use a 5 mL syringe to squeeze a small amount of warm gelatin into the bottom of the cup and allow to cool slightly until it becomes sticky. This will assist in keeping the tissue stuck to the bottom of the cup and prohibit it from floating in the gelatin.

4. Keep gelatin slightly warm before pouring over tissue. Too warm gelatin will cause the gelatin in the bottom of the cup to liquefy and the tissue to float.

5. It is a good idea to mark the cryostat cup and sketch the orientation of the tissue in your notebook for future reference. Note which side of the tissue is facing the bottom of the cup (tissue slicing begins from the bottom of the cup).

6. Use large forceps or another apparatus to hold the cup upright. Be sure not to get ethanol into the cup; this will cause the gelatin to dissolve and not freeze completely.

7. Due to the interference caused by OCT media, OCT is used only to mount the gelatin to the chuck, but should not touch the tissue sample itself.

8. The cryostat is most commonly kept at –20 °C for tissue sectioning; however setting the temperature to –25 °C results in better slicing for some types of tissues.

9. For larger tissue samples such as whole-body animal samples, use a section thickness of 20 μm.

10. There are two methods for thaw mounting. In the first method, warm an ITO-coated glass slide by placing the back of your hand on the back of the slide until warm. Place the ITO-coated side of the warmed slide near the cold slice and allow the slice to melt onto the slide. For the second method, use a fine paint-brush to place the cold slice onto the cold slide and warm the slide and slice together with the back of your hand as mentioned previously. The latter method is trickier and can ruin the section, but also results in less analyte loss [22].

11. It is beneficial to check the integrity of the tissue sections by thaw mounting the first section onto a regular glass microscope slide and carefully examining it under a microscope before placing sections onto the more expensive ITO-coated glass slides.

12. Different organs from different biological tissues have different thicknesses, so the number of slices and the distance between slices necessary for 3D imaging vary.

13. Use the airbrush in a fume hood to avoid inhaling matrix solution. Hold the airbrush perpendicular to the glass slide, 35 cm away, and adjust the flow rate so that most of the matrix solvent evaporates before reaching the slide.

14. HTX Technologies has recommended methods that can be used as a starting point. Changing variables such as flow rate, velocity, temperature, and number of passes will change the dryness of the spray, the coverage, and the amount of matrix deposited. For MSI of metabolites in root nodules using 40 mg/mL DHB as the matrix, set the temperature to ~80 °C, velocity to 1,250 mm/min, flow rate to 50 µL/min, and number of passes to 24. For best coverage, it is recommended to rotate the nozzle 90° and/or offset the nozzle 1.5 mm between each pass. When using other matrices be sure to use the same solvent in the syringe pump that the matrix is dissolved in. The temperature should also be adjusted when using other matrices; slowly increase the temperature until you hear a "puffing" sound and then reduce the temperature by 5 °C for your final method.

15. If the glass slide is too large for the sublimation chamber, cut the glass slide to the appropriate size (i.e., cut in half) and apply the matrix to both halves separately. Take this into account when placing cryostat tissue slices onto the slide.

16. With EPSON Scan software, hit the "Preview" button and draw a box around the image of the slide. Hit the "Zoom" button to zoom in on the slide region. Set the resolution to 2,400 dpi and scan the image. Open the image and orient the picture to match the orientation of the slide in the holder.

17. To build an acquisition file, experiment with different laser diameters, laser intensities, and number of shots to optimize

the signal intensity. Typically use a raster width of 50 μm; smaller raster widths give higher resolution images but takes longer time to acquire images. "LIFT" mode allows the acquisition of MS/MS spectra.

18. The WiteOut marks can be used as teach points. After setting the teach points, select the "move sample carrier" option on flexImaging and move the plate around by clicking on different areas of the optical image to make sure that the alignment is acceptable.

19. To calibrate with matrix peaks, shoot the laser at a spot of matrix only.

20. If you are using the airbrush matrix application method, it is beneficial to image several spots of pure matrix in case there are inconsistencies with the spray coverage.

21. Several other 3D imaging software packages can also be used, such as LSM Viewer (Zeiss, Germany), Metamorph (Molecular Devices, Sunnyvale, CA, USA), and Amira (Mercury Computer Systems, Chelmsford, MA, USA). Other commercial, open-source 3D imaging software platforms include Fiji, CellProfiler, Vaa3D, BioImageXD, Icy, and Konstanz Information Miner [23].

22. The images must be in gray scale to be processed with Image J.

Acknowledgments

This work was supported by funding from the University of Wisconsin Graduate School and the Wisconsin Alumni Research Foundation (WARF) and Romnes Faculty Research Fellowship program (to L.L.). E.G. acknowledges a National Science Foundation (NSF) Graduate Research Fellowship. (DGE-1256259). The acquisition of the TM sprayer was funded by an NIH shared instrument grant 1S10RR029531.

References

1. Singh RJ, Chung GH, Nelson RL (2007) Landmark research in legumes. Genome 50(6):525–537. doi:10.1139/g07-037

2. Barsch A, Patschkowski T, Niehaus K (2004) Comprehensive metabolite profiling of Sinorhizobium meliloti using gas chromatography–mass spectrometry. Funct Integr Genomics 4(4):219–230. doi:10.1007/s10142-004-0117-y

3. Desbrosses GG, Kopka J, Udvardi MK (2005) Lotus japonicus metabolic profiling. Development of gas chromatography–mass spectrometry resources for the study of plant-microbe interactions. Plant Physiol 137(4):1302–1318. doi:10.1104/pp.104.054957

4. Prell J, Poole P (2006) Metabolic changes of rhizobia in legume nodules. Trends Microbiol 14(4):161–168. doi:10.1016/j.tim.2006.02.005

5. Kutz KK, Schmidt JJ, Li LJ (2004) In situ tissue analysis of neuropeptides by MALDI FTMS in-cell accumulation. Anal Chem 76(19):5630–5640. doi:10.1021/Ac049255b

6. Stemmler EA, Cashman CR, Messinger DI, Gardner NP, Dickinson PS, Christie AE (2007) High-mass-resolution direct-tissue MALDI-FTMS reveals broad conservation of

three neuropeptides (APSGFLGMRamide, GYRKPPFNGSIFamide and pQDLDHVFL-RFamide) across members of seven decapod crustacean infraorders. Peptides 28(11):2104–2115. doi:10.1016/j.peptides.2007.08.019

7. Rubakhin SS, Churchill JD, Greenough WT, Sweedler JV (2006) Profiling signaling peptides in single mammalian cells using mass spectrometry. Anal Chem 78(20):7267–7272. doi:10.1021/Ac0607010

8. Neupert S, Predel R (2005) Mass spectrometric analysis of single identified neurons of an insect. Biochem Biophys Res Commun 327(3):640–645. doi:10.1016/j.bbrc.2004.12.086

9. Cornett DS, Reyzer ML, Chaurand P, Caprioli RM (2007) MALDI imaging mass spectrometry: molecular snapshots of biochemical systems. Nat Methods 4(10):828–833. doi:10.1038/Nmeth1094

10. Chaurand P, Norris JL, Cornett DS, Mobley JA, Caprioli RM (2006) New developments in profiling and imaging of proteins from tissue sections by MALDI mass spectrometry. J Proteome Res 5(11):2889–2900. doi:10.1021/Pr060346u

11. Svensson M, Boren M, Skold K, Falth M, Sjogren B, Andersson M, Svenningsson P, Andren PE (2009) Heat stabilization of the tissue proteome: a new technology for improved proteomics. J Proteome Res 8(2):974–981. doi:10.1021/Pr8006446

12. Caprioli RM, Farmer TB, Gile J (1997) Molecular imaging of biological samples: localization of peptides and proteins using MALDI-TOF MS. Anal Chem 69(23):4751–4760

13. Ye H, Gemperline E, Li L (2013) A vision for better health: mass spectrometry imaging for clinical diagnostics. Clin Chim Acta 420:11–22. doi:10.1016/j.cca.2012.10.018

14. Thomas A, Charbonneau JL, Fournaise E, Chaurand P (2012) Sublimation of new matrix candidates for high spatial resolution imaging mass spectrometry of lipids: enhanced information in both positive and negative polarities after 1,5-diaminonaphthalene deposition. Anal Chem 84(4):2048–2054. doi:10.1021/ac2033547

15. Shroff R, Rulisek L, Doubsky J, Svatos A (2009) Acid–base-driven matrix-assisted mass spectrometry for targeted metabolomics. Proc Natl Acad Sci U S A 106(25):10092–10096. doi:10.1073/pnas.0900914106

16. Chen S, Chen L, Wang J, Hou J, He Q, Liu J, Xiong S, Yang G, Nie Z (2012) 2,3,4,5-Tetrakis(3′,4′-dihydroxylphenyl)thiophene: a new matrix for the selective analysis of low molecular weight amines and direct determination of creatinine in urine by MALDI-TOF MS. Anal Chem 84(23):10291–10297. doi:10.1021/ac3021278

17. Shrivas K, Hayasaka T, Sugiura Y, Setou M (2011) Method for simultaneous imaging of endogenous low molecular weight metabolites in mouse brain using TiO2 nanoparticles in nanoparticle-assisted laser desorption/ionization-imaging mass spectrometry. Anal Chem 83(19):7283–7289. doi:10.1021/ac201602s

18. Baluya DL, Garrett TJ, Yost RA (2007) Automated MALDI matrix deposition method with inkjet printing for imaging mass spectrometry. Anal Chem 79(17):6862–6867. doi:10.1021/ac070958d

19. Hankin JA, Barkley RM, Murphy RC (2007) Sublimation as a method of matrix application for mass spectrometric imaging. J Am Soc Mass Spectrom 18(9):1646–1652. doi:10.1016/j.jasms.2007.06.010

20. Ye H, Gemperline E, Venkateshwaran M, Chen R, Delaux PM, Howes-Podoll M, Ane JM, Li L (2013) MALDI mass spectrometry-assisted molecular imaging of metabolites during nitrogen fixation in the *Medicago truncatula-Sinorhizobium meliloti* symbiosis. Plant J 75(1):130–145. doi:10.1111/tpj.12191

21. Robichaud G, Garrard KP, Barry JA, Muddiman DC (2013) MSiReader: an open-source interface to view and analyze high resolving power MS imaging files on Matlab platform. J Am Soc Mass Spectrom 24(5):718–721. doi:10.1007/s13361-013-0607-z

22. Schwartz SA, Reyzer ML, Caprioli RM (2003) Direct tissue analysis using matrix-assisted laser desorption/ionization mass spectrometry: practical aspects of sample preparation. J Mass Spectrom 38(7):699–708. doi:10.1002/Jms.505

23. Schindelin J, Arganda-Carreras I, Frise E, Kaynig V, Longair M, Pietzsch T, Preibisch S, Rueden C, Saalfeld S, Schmid B, Tinevez JY, White DJ, Hartenstein V, Eliceiri K, Tomancak P, Cardona A (2012) Fiji: an open-source platform for biological-image analysis. Nat Methods 9(7):676–682. doi:10.1038/Nmeth.2019

Chapter 5

MALDI Mass Spectrometry Imaging of Lipids and Primary Metabolites on Rat Brain Sections

David Touboul and Alain Brunelle

Abstract

Matrix-assisted laser desorption/ionization mass spectrometry imaging (MALDI MSI) enables the localization and the structural identification of a large set of molecules on a surface with 10–20 μm resolution. MALDI is often coupled to time-of-flight (TOF) tandem analyzers which remain a versatile instrument for the detection and the structural analysis of molecular ions. This technique can be used to locate either large biomolecules, such as peptides/proteins, or small endogenous or exogenous ones. Among them, lipids and primary metabolites are of high interest because they can reflect the cell state. This chapter is thus dedicated to the analysis, on rat brain tissue sections, of lipids in positive and negative ion modes, and primary metabolites in the negative ion mode. A particular attention is paid to the structural characterization of lipids using lithium cationization in the positive ion mode.

Key words Matrix-assisted laser desorption/ionization, Time-of-fight, Mass spectrometry imaging, Tandem mass spectrometry, Lipid, Primary metabolite

1 Introduction

Mass spectrometry imaging (MSI) has gained importance since the last 15 years. Matrix-assisted laser desorption/ionization (MALDI) MSI was firstly described in 1994 by Spengler [1] and by the group of Caprioli in 1997 for the localization of peptides and proteins on tissue surface [2, 3]. Since these first demonstrations, the technique was hardly improved in terms of robustness, sensitivity, spatial resolution, and data analysis. Nevertheless, the identification of the peptide/protein peaks in the m/z range between 1,500 and 25,000 remains highly challenging and requires either unsensitive top-down approaches [4] or extraction/digestion/separation steps [5]. In parallel, MALDI MSI was also used for the localization of lipid [6, 7] and primary metabolite [8] ion species in the m/z range between 250 and 1,500. In fact lipids are major constituents of cells and partially reflect the tissue state. Moreover, due to the development of instruments, such as time of flight (TOF) [9] or

Lin He (ed.), *Mass Spectrometry Imaging of Small Molecules*, Methods in Molecular Biology, vol. 1203, DOI 10.1007/978-1-4939-1357-2_5, © Springer Science+Business Media New York 2015

Orbitrap™ [10], exact mass measurements can be performed directly on tissue sections. MS/MS capabilities of most of the mass spectrometers dedicated to MSI allow the rapid structural identification of the lipid species without requiring extraction or separation.

In this chapter sample preparation for MALDI MSI of lipids [11, 12] and primary metabolites [8] on rat brain sections is described in positive and negative ion modes. In fact rat brain can be considered as a good model sample due to its high availability, its easy sample preparation, and its lipid composition depending on the different anatomical areas. Moreover, lithium cationization of the lipid species leads to better structural characterization in the positive ion mode [13]. For matrix deposition, several automated technologies are commercially available. Most of them are based on the nebulization of the matrix solution on the tissue section using a robot and allow reproducible sample preparation. Because of its ease of use and robustness, the methods presented here were developed using the TM-Sprayer™ from HTX-Imaging.

2 Materials

2.1 Chemicals

1. α-Cyano-4-hydroxycinnamic acid (CHCA): 10 mg/mL, in acetonitrile/water/trifluoroacetic acid (70/30/0.1, $v/v/v$).

2. 9-Aminoacridine (9-AA): 10 mg/mL in ethanol/water (70/30, v/v).

3. Lithium trifluoroacetate.

4. Trifluoroacetic acid.

5. HPLC-grade water.

6. Ethanol.

7. Acetonitrile.

8. Sodium pentobarbital.

9. Optimal cutting temperature (OCT) embedding medium.

10. Dry ice.

2.2 Instruments and Materials

1. TM-Sprayer™ from HTX-Imaging (Carrboro, NC, USA) or equivalent robot for matrix deposition.

2. 4800 MALDI TOF/TOF mass spectrometer from AB Sciex (Les Ulis, France) equipped with 4000 Series Imaging software (www.maldi-msi.org, M. Stoeckli, Novartis Pharma, Basel, Switzerland) and Tissue View software (AB Sciex, Les Ulis, France) or equivalent.

3. Cryostat (model CM3050-S; Leica Microsystems SA, Nanterre, France, or equivalent).

4. Optical microscope (Olympus BX 51 fitted with ×1.25 to ×50 lenses) (Olympus France SAS, Rungis, France) equipped with a Color View I camera, monitored by Cell[B] software (Soft Imaging System GmbH, Münster, Germany) or equivalent.

5. Desiccator.

6. Sonicator device.

7. Vortex mixer.

8. Sample plate for MALDI instrument (conductive glass slide coated with indium tin oxide or stainless steel plate).

3 Methods

3.1 Tissue Sectioning

All experiments on rat brains were performed in accordance with the protocols approved by the National Commission on animal experimentation and by the recommendations of the European commission DGXI.

1. Euthanize male Wistar rats of typical 400 g weight by an intraperitoneal injection of sodium pentobarbital (>65 mg/kg).

2. Freeze the trimmed tissue blocks immediately in dry ice to prevent crack formation during freezing and store at –80 °C prior to MS experiments (*see* **Note 1**).

3. Add a few drops of OCT embedding medium and quickly deposit the tissue block over it to fix the frozen tissue block to the cryostat adapter (*see* **Note 2**).

4. Cut tissue section of 12 μm thickness using the cryostat at –20 °C.

5. Deposit two successive tissue sections on either a conductive glass slide or stainless plate adapted to the MALDI holder (*see* **Notes 3** and **4**).

6. Store the samples at –80 °C in a box to prevent degradation by ice.

7. Before MS analysis, warm the sample at room temperature for a few seconds and dry it under vacuum at a pressure of a few hPa for 30 min in a desiccator. Calibrants can be deposited on the tissue section required for the instrumental optimization.

8. Take pictures of the sample using a microscope (*see* **Note 5**).

3.2 Lipid MALDI MSI in the Positive Ion Mode

1. Prepare 10 mL of the CHCA matrix solution. The solution needs to be vortexed for 1 min and then sonicated for 5 min before being sprayed.

2. Fill the reservoir loop of the TM-Sprayer™. Five milliliter is enough for the matrix deposition over four different sample plates.

3. Fix the flow rate of the isocratic pump at 240 μL/min, the temperature of the nozzle/air spray at 120 °C, and the velocity of the sample stage at 120 cm/min.

4. Verify that the nozzle is not blocked and that the matrix is correctly deposited using a blank plate. If the matrix surface looks homogeneous, place the sample plate and deposit the matrix. The matrix deposition is done in less than 2 min. Rinse the complete system with an acetonitrile/water/trifluoroacetic acid (70/30/0.1, $v/v/v$) mixture for 30 min.

5. Introduce the sample plate in the vacuum chamber of the mass spectrometer. The time needed to reach the pressure limit allowing the plate admission in the analysis chamber can be longer than for classical dried-droplet MALDI sample.

6. A first tissue section will be used to optimize the instrumental parameters, i.e., the laser intensity, the extraction delay, and the number of laser shots. Laser intensity is usually fixed 10 % higher than the ionization threshold whereas a typical value for the extraction delay is 450 ns. The number of laser shot is chosen by firing the sample at a fixed position and monitoring the signal-to-noise ratio of the lipid signals. Typical number of laser shots per pixel is about 150.

7. Calibrate the mass spectrometer. Two calibration methods can be chosen: the first consists in the deposition of calibrants at a higher concentration on the tissue section before the matrix deposition. In that case, the calibrant ions are generated at the same position as the lipid species and thus a 10 ppm internal calibration can be achieved. The second method consists in using the signal of known lipid species, for example the ions corresponding to three glycerophosphatidylcholines (PCs) at m/z 758.5700 ([PC 34:2+H]$^+$), m/z 782.5676 ([PC34:1+Na]$^+$), and m/z 798.5416 ([PC34:1+K]$^+$).

8. Determine the tissue area which will be imaged on the second tissue section.

9. Choose the pixel size. Minimal pixel size can be determined by the dimension of the crater formed after 150 laser shots at a fixed position on the tissue sample. Depending on the instrument, pixel size can be from 5 to 100 μm (*see* **Note 6**). Oversampling can also be considered. In that the pixel size is smaller than the crater size. Sensitivity is usually lower when using this acquisition mode.

10. Start the acquisition. The total acquisition time is calculated by the 4000 Series Imaging software. Typical acquisition time per pixel is between 1 and 2 s depending on the number of laser shots per pixel.

11. Remove the sample from the mass spectrometer. The sample can be kept for several days in a desiccator without significant degradation. It can be useful to keep the sample if structural characterization of lipid species is required.

12. Analyze the data using Tissue View software or equivalent ones. A list of available data treatment software is provided on the website http://www.maldi-msi.org/. Typical brain lipid images are presented in literature [11]. Similar lipid distribution should be obtained following the above procedure.

3.3 Structural Characterization of Lipids Using Lithium Cationization in the Positive Ion Mode

1. Prepare 10 mL of the CHCA matrix solution. Add a solution of lithium trifluoroacetate at 2 mg/mL (1/1, v/v). The final solution needs to be vortexed during 1 min and then sonicated during 5 min before being sprayed.

2. Fill the reservoir loop of the TM-Sprayer™. Five milliliter is enough for the matrix deposition over four different sample plates.

3. Fix the flow rate of the isocratic pump at 300 µL/min, the temperature of the nozzle/air spray at 150 °C, and the velocity of the sample stage at 120 cm/min.

4. Verify that the nozzle is not blocked and that the matrix is correctly deposited using a blank plate. If the matrix surface looks homogeneous, place the sample plate and deposit the matrix. The matrix deposition is done in less than 2 min. Rinse the complete system with an acetonitrile/water/trifluoroacetic acid (70/30/0.1, $v/v/v$) mixture for 30 min.

5. Introduce the sample plate in the vacuum chamber of the mass spectrometer. The time needed to reach the pressure limit for introducing the sample in the analysis chamber can be longer than for classical dried-droplet MALDI sample.

6. A first tissue section will be used to optimize the instrumental parameters, i.e., the laser intensity, the extraction delay, and the number of laser shots. Laser intensity is usually fixed 10 % higher than the ionization threshold whereas a typical value for the extraction delay is 450 ns. The number of laser shot is chosen by firing the sample at a fixed position and monitoring the signal-to-noise ratio of the lipid signals. Typical number of laser shots per pixel is about 150.

7. Calibrate the MS mode. In the case of lithium adduct, calibration must be done with known lipid signals such as those at m/z 740.5782 ([PC 32:0 + Li]$^+$) and m/z 766.5938 ([PC 34:1 + Li]$^+$).

8. Calibrate the MS/MS mode using specific fragments of the ion at m/z 766.5938 ([PC 34:1 + Li]$^+$): 707.6 (loss of triethylamine), 583.6 (loss of phosphocholine), 577.6 (loss of phos-

phocholine + lithium), 190.2 (phosphocholine cationized by lithium), and 86.1 (dehydroxycholine) [13].

9. Acquire MS/MS spectra of lithium-cationized lipids using the "metastable suppressor ON" mode. Selection window need to be optimized in order to select a single lipid species without a too significant loss of sensitivity. Spectra can be compared to those previously published [13].

10. If MALDI MS/MS imaging is required, same procedure as described in Subheading 3.2 can be followed.

3.4 Lipid and Primary Metabolite MALDI MSI in the Negative Ion Mode

1. For lipid analysis, prepare 10 mL of the 9-AA matrix solution. For primary metabolite analysis, add 1 % of trifluoroacetic acid to the matrix solution. The solution needs to be vortexed during 1 min and then sonicated during 5 min before being sprayed.

2. Fill the reservoir loop of the TM-Sprayer™. Five milliliter is enough for the matrix deposition over four different sample plates.

3. Fix the flow rate of the isocratic pump at 240 µL/min, the temperature of the nozzle/air spray at 80 °C, and the velocity of the sample stage at 120 cm/min.

4. Verify that the nozzle is not blocked and that the matrix is correctly deposited using a blank plate. If the matrix surface looks homogeneous, place the sample plate and deposit the matrix. The matrix deposition is done in less than 2 min. Rinse the complete system with an ethanol/water (70/30, v/v) mixture for 30 min.

5. Introduce the sample plate in the vacuum chamber of the mass spectrometer. The time needed to reach the pressure limit for introducing the sample in the analysis chamber can be longer than for classical dried-droplet MALDI sample.

6. A first tissue section will be used to optimize the instrumental parameters, i.e., the laser intensity, the extraction delay, and the number of laser shots. Laser intensity is usually fixed 20 % higher than the ionization threshold whereas a typical value for the extraction delay is 450 ns. The number of laser shot is chosen by firing the sample at a fixed position and monitoring the signal-to-noise ratio of the lipid signals. Typical number of laser shots per pixel is about 200 for lipid analysis and 80 for primary metabolite analysis.

7. Calibrate the mass spectrometer. Calibration must be done with known lipid signals such as those at m/z 700.5281 (plasmalogen phosphatidylethanolamine 34:1, [PE-p34:1-H]$^-$), m/z 788.5442 (glycerophosphatidylserine 36:1, [PS 36:1-H]$^-$), and m/z 888.62351 (sulfatide C24:1, [ST 24:1-H]$^-$) or with known primary metabolite signals such as those at m/z

346.0553 ([adenosine monophosphate -H]⁻), m/z 426.02164 ([adenosine diphosphate -H]⁻), and m/z 505.98799 ([adenosine triphosphate -H]⁻).

8. Calibrate the MS/MS mode using the previously described fragments of either the ion at m/z 888.62351 (sulfatide C24:1, [ST 24:1-H]⁻) or the ion at m/z 427.02947 ([adenosine diphosphate -H]⁻).

9. Acquire MS images or MS/MS spectra as described above in Subheadings 3.2 and 3.3.

4 Notes

1. Formaldehyde fixation is always avoided when possible in order not to chemically modify the lipid species [14].

2. The transparent OCT embedding medium slowly becomes white when hardens.

3. One tissue section is needed for the optimization of the instrumental parameter whereas a second one is for the MALDI MSI experiment.

4. Classical histochemistry (hematoxylin and eosin staining for example) is usually performed on adjacent tissue sections in order to visualize the different anatomical parts of the rat brain. It must be noticed that histochemistry can also be performed on the same tissue section that has been used for MALDI MSI experiments [15, 16].

5. It is important to keep the same orientation of the sample between the microscope observation and the MALDI MSI acquisition in order to easily correlate the images.

6. 4800 MALDI TOF/TOF (AB Sciex) works with a fixed laser focus. It must be noted that high value of laser power is related to larger amount of ablated material, i.e., larger crater size. Other MALDI instruments allow the modification of the laser focus depending on the applications (e.g., proteomics for which high sensitivity is required versus MSI for which high lateral resolution is required).

References

1. Spengler B, Hubert M, Kaufmann R (1994) In: Proceedings of the 42nd Annual Conference on Mass Spectrometry and Allied Topics, Chicago, IL, p 1041

2. Caprioli RM, Farmer TB, Gile J (1997) Molecular imaging of biological samples: localization of peptides and proteins using MALDI-TOF MS. Anal Chem 69(23): 4751–4760

3. Stoeckli M, Chaurand P, Hallahan DE et al (2001) Imaging mass spectrometry: a new technology for the analysis of protein expression in mammalian tissues. Nat Med 7(4): 493–496

4. Debois D, Smargiasso N, Demeure K et al (2013) MALDI in-source decay, from sequencing to imaging. Top Curr Chem 331: 117–141

5. Maier SK, Hahne H, Moghaddas Gholami A et al (2013) Comprehensive identification of proteins from MALDI imaging. Mol Cell Proteomics 12:2901–2910. doi:10.1074/mcp.M113.027599

6. Touboul D, Laprévote O, Brunelle A (2011) Micrometric molecular histology of lipids by mass spectrometry imaging. Curr Opin Chem Biol 15(5):725–732

7. Fernández JA, Ochoa B, Fresnedo O et al (2011) Matrix-assisted laser desorption ionization imaging mass spectrometry in lipidomics. Anal Bioanal Chem 401(1):29–51

8. Benabdellah F, Touboul D, Brunelle A et al (2009) In situ primary metabolites localization on a rat brain section by chemical mass spectrometry imaging. Anal Chem 81(13):5557–5560

9. Guilhaus M (1995) Principles and instrumentation in time-of-flight mass spectrometry. Physical and instrumental concepts. J Mass Spectrom 30(11):1519–151532

10. Zubarev RA, Makarov A (2013) Orbitrap mass spectrometry. Anal Chem 85(11):5288–5296

11. Benabdellah F, Seyer A, Quinton L et al (2010) Mass spectrometry imaging of rat brain sections: nanomolar sensitivity with MALDI versus nanometer resolution by TOF-SIMS. Anal Bioanal Chem 396(1):151–162

12. Cerruti CD, Benabdellah F, Laprévote O et al (2012) MALDI imaging and structural analysis of rat brain lipid negative ions with 9-aminoacridine matrix. Anal Chem 84(5):2164–2171

13. Cerruti CD, Touboul D, Guérineau V et al (2011) MALDI imaging mass spectrometry of lipids by adding lithium salts to the matrix solution. Anal Bioanal Chem 401(1):75–87

14. Eltoum I, Fredenburgh J, Myers RB et al (2001) Introduction to the theory and practice of fixation of tissues. J Histotechnol 3(18):173–190

15. Schwamborn K, Krieg RC, Reska M et al (2007) Identifying prostate carcinoma by MALDI-Imaging. Int J Mol Med 20(2):155–159

16. Deutskens F, Yang J, Caprioli RM (2011) High spatial resolution imaging mass spectrometry and classical histology on a single tissue section. J Mass Spectrom 46(6):568–571

Chapter 6

Multiplex MALDI-MS Imaging of Plant Metabolites Using a Hybrid MS System

Andrew R. Korte, Gargey B. Yagnik, Adam D. Feenstra, and Young Jin Lee

Abstract

Plant tissues present intriguing systems for study by mass spectrometry imaging, as they exhibit a complex metabolism and a high degree of spatial localization. This chapter presents a methodology for preparation of plant tissue sections for matrix-assisted laser desorption/ionization mass spectrometry imaging (MALDI-MSI) analysis and the use of a hybrid mass spectrometer for "multiplex" imaging. The multiplex method described here provides a wide range of analytical information, including high-resolution, accurate mass imaging and tandem MS scans for structural information, all within a single experiment. While this procedure was developed for plant tissues, it can be readily adapted for analysis of other sample types.

Key words Mass spectrometry imaging, Multiplex imaging, Plant metabolites, Hybrid MS, Orbitrap

1 Introduction

Matrix-assisted laser desorption/ionization-mass spectrometry imaging (MALDI-MSI) has been extensively utilized for analysis of animal and human tissues, but it is also a promising technique for analysis of plant metabolites [1, 2]. Plant metabolism is complex; the number of unique metabolites in the plant kingdom has been estimated as high as 200,000 [3]. It is also highly localized, with many specialized structures and tissues and significant variation in metabolic profiles—even between, for example, the organs of a flower [4]. MALDI-MSI, with its ability to provide rich chemical information in high spatial resolution, is beginning to be applied to plant systems for metabolite analysis. Efforts so far include mapping of lipids in cottonseeds [5] and on plant surfaces [6, 7], sugars in wheat seeds [8], cellulosic carbohydrates in wood [9], and secondary metabolites in flower petals [10]. Plant applications for MS imaging using other ionization techniques have also

Lin He (ed.), *Mass Spectrometry Imaging of Small Molecules*, Methods in Molecular Biology, vol. 1203,
DOI 10.1007/978-1-4939-1357-2_6, © Springer Science+Business Media New York 2015

been reported, such as laser ablation electrospray ionization [11] and desorption electrospray ionization [12].

The development of hybrid mass spectrometers, which incorporate more than one type of mass analyzer into a single instrument, has significantly expanded the capabilities of mass spectrometry for metabolomic analysis. Existing mass analyzers generally compromise one or more of the following: scan speed, mass resolution, and ability for tandem MS scans [2]. Combining two complementary mass analyzers into a single analytical platform helps to overcome some of these limitations. One example of this kind of instrumentation is a linear ion trap-orbitrap, which incorporates a linear ion trap for high scan rates and MS^n scans, and an orbitrap analyzer for high mass resolution and accurate mass measurements. Using this instrumentation, we developed a "multiplex" MS imaging technique to perform high-mass-resolution imaging, MS/MS and/or MS^n imaging, and data-dependent tandem MS scans on a single tissue section within a single experiment [10, 13]. This method yields a wealth of analytically useful information, such as accurate mass measurements for assignment of molecular formulas, spatial discrimination of structural isomers, and structurally informative MS/MS spectra. While our method has been developed for a linear ion trap-orbitrap mass spectrometer, the procedure could be adapted for other tandem MS-capable systems.

Here we describe a sample preparation method for MSI, developed to maintain high spatial localization of plant metabolites down to the single-cell (~10 μm) level. We also present an instrumental methodology for multiplex MS imaging to obtain rich chemical information in a single imaging experiment. Fig. 1 illustrates the overall procedure from tissue harvest to data analysis with each corresponding section number. The reader can refer to specific sections of their interest. Fig. 2 illustrates an example of the protocol introduced in this chapter applied to a germinating corn seed.

2   Materials

2.1   Tissue Cryosectioning

1. Cryomicrotome: A cryomicrotome consists of a microtome inside a cryostat, allowing for the cutting of thin sections of frozen tissue. Precool the cryostat to the appropriate temperature, e.g., –20 °C, and keep microscope slides and adhesive tape sections inside the cryostat.

2. High-purity water: Nanopure or LC-MS-grade water is recommended to minimize any mass spectrometry contamination.

3. Warm gelatin solution: Prepare a 10 % w/v solution (e.g., 1 g in 10 mL) of 300 bloom gelatin in water by heating the water to ~75 °C, then adding the gelatin, and stirring manually until dissolved.

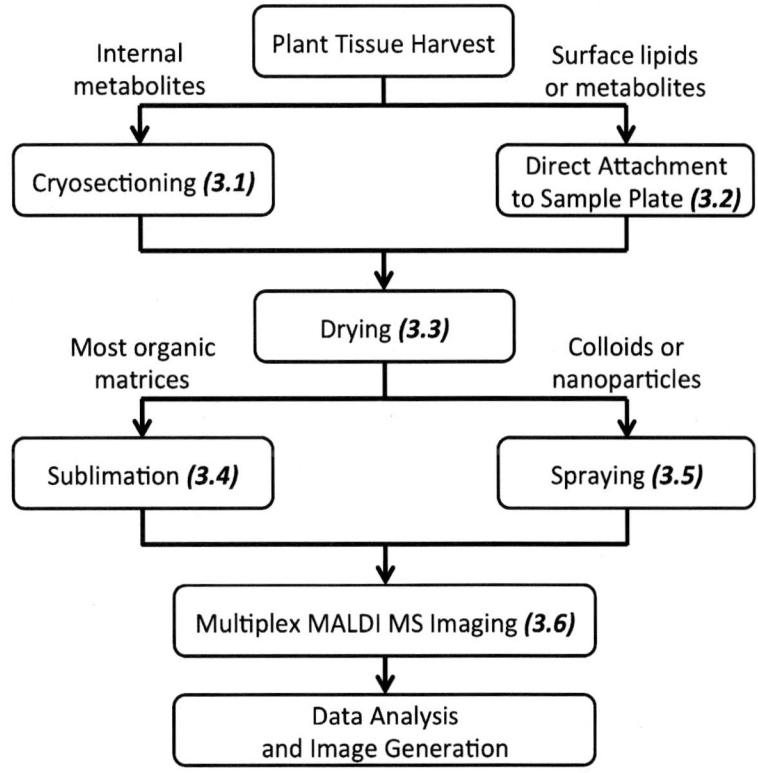

Fig. 1 Workflow for multiplex MALDI-MS imaging of plant metabolites. Procedures discussed in this work are labeled with the corresponding section number

4. Cryomold: This should be large enough to contain tissue sample and gelatin-embedding medium.

5. Liquid nitrogen in a dewar: Pour into a Styrofoam box just before flash-freezing tissue or cryomold.

6. 70 % ethanol solution, prepared from LC-MS-grade ethanol and water.

7. Cryo-Jane® adhesive tape sections (Leica Biosystems, Buffalo Grove, IL, USA; *see* **Note 1**).

8. Optimal cutting temperature (OCT) compound.

9. Glass microscope slides: These slides are to carry and store tissue sections.

10. Styrofoam cooler with dry ice.

2.2 Direct Attachment of Intact Tissues

1. Double-sided tape.

2. Sample-handling tools (e.g., forceps).

3. Tank of compressed nitrogen.

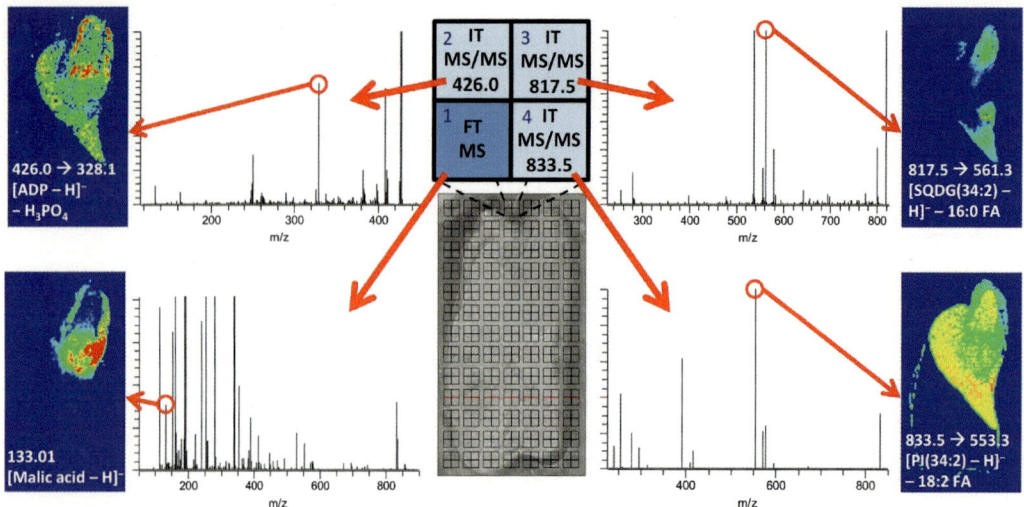

Fig. 2 Illustration of the protocol applied to a germinating corn seed. A corn seed was germinated in water for 3 days in a greenhouse, cryosectioned (Subheading 3.1), freeze-dried (Subheading 3.3), and sublimated with 9-aminoacridine matrix (Subheading 3.4), before a four-step multiplex MS imaging experiment was performed (raster design A in Fig. 3). ADP, SQDG, and PI represent adenosine diphosphate, sulfoquinovosyl diacylglycerol, and phosphatidylinositol, respectively. Peaks used to generate images are *circled*

2.3 Sample Drying

1. Roll of adhesive tape (e.g., electrical tape): To attach the tissue sections to a heat sink.

2. Vacuum chamber or lyophilizer with vacuum system capable of mtorr pressures.

3. Heat sink (e.g., metal block; *see* **Note 2**): Precool the heat sink in a −80 °C freezer for several hours.

2.4 Matrix Application by Sublimation

1. Sublimation apparatus assembly: A sublimation apparatus with a flat condenser bottom, a cold trap (*see* **Note 3**), a pressure gauge, and a rotary vacuum pump (*see* the supplemental information of [14] for a detailed schematic of a comparable setup).

2. Glass microscope slide (optional): To attach the tissue slice to a smaller size sublimation device. Cut into half with glass scorer. Mounting Cryo-Jane tape windows with tissue sections to a glass slide makes handling and removing the sample easier.

3. Heating assembly: Temperature-controllable heating mantle and controller. The mantle should be preheated to the intended temperature.

4. Roll of adhesive tape.

5. Crushed dry ice.

6. Acetone (any grade).

7. Appropriate MALDI matrix (*see* **Note 4**).

2.5 Matrix Application by Oscillating Capillary Nebulizer

1. Matrix solution/suspension (*see* **Note 5**).

2. 500 µL syringe: Rinse the syringe with the same solvent used for matrix.

3. Syringe pump.

4. Oscillating capillary nebulizer (OCN): A handheld airbrush or other commercial nebulizer can also be used if the matrix homogeneity is not a concern. We use an OCN to ensure homogeneous matrix application ($\leq$ ~10 µm). One can make such a device by simple modification of a commercial airbrush (Aztek A470; Testor, Rockford, IL) (Fig. 2 and *see* **Note 6**). Rinse the capillary with the same solvent used for matrix.

5. Tank of compressed nitrogen.

3 Methods

3.1 Tissue Sectioning

This section is intended for imaging of internal metabolites with minimal analyte loss or redistribution.

1. Harvest the tissue from the plant, and flash-freeze it as quickly as possible by submerging it into liquid nitrogen. Keep the frozen tissue in a cooler with dry ice while transporting it to the cryostat.

2. Place the frozen tissue into the precooled cryostat for approximately 30 min to allow it to warm to the temperature of the cryostat.

3. Place the frozen tissue into the mold, making sure to orient the tissue so as to section in the desired plane. Pour the warm gelatin solution around the tissue to fill the mold.

4. Float the mold on liquid nitrogen until the gelatin is almost completely frozen to the center (~10–20 s), and then transfer the mold into the cryostat (*see* **Note 7**). Once the gelatin is completely frozen (*see* **Note 8**), let it stay in the cryostat for an additional 30 min to ensure that the tissue block has equilibrated to the temperature of the cryostat.

5. Remove the tissue block from the mold by cutting the sides of the mold with a razor blade and carefully peeling away the plastic of the cryomold.

6. Place a small amount (<0.5 mL) of OCT compound on the cryotome sample stage and immediately press the tissue block onto it. Allow the OCT to set and fix the tissue block to the stage (*see* **Note 9**).

7. Before installing the cryomicrotome blade, rinse it several times with 70 % ethanol to remove any oil or other contaminants that may be transferred to the tissue during sectioning.

8. Run off several sections to provide a flat sample surface and reach the desired portion of the embedded tissue. Set the tissue thickness to the desired value (e.g., 10–20 μm).

9. Remove the protective strip from the Cryo-Jane tape and stick the tape window to the tissue surface (*see* **Note 10**). Using a roller or similar tool, carefully press the tape against the tissue section for uniform adhesion.

10. Slice the section using the cryomicrotome.

11. Keeping it deep inside the cryostat, collect the tape with attached section and inspect it for any potential damage during cutting. If the sample is damaged, discard it and repeat **steps 9** and **10**.

12. Place the tape window with section attached face up on a chilled glass slide, and attach it by taping both ends to the slide. Ensure that the tape window is flat against the slide (*see* **Note 11**). Avoid prolonged contact with either the tape or slide to prevent thawing of the sample.

13. Remove the slide, with the tape and section attached, from the cryostat and quickly transfer them into a covered cooler full of dry ice (*see* **Note 12**).

14. Repeat **steps 9–13** until the desired number of sections has been collected.

15. Optionally, some sections may be taken using traditional thaw-mounting for imaging with optical microscopy, with possible fixation and/or staining.

16. Store tissue sections at −80 °C until analysis.

3.2 Direct Attachment to Sample Plate

This section is intended for imaging of surface metabolites.

1. Place a strip or several strips of double-sided tape on the MALDI sample plate. Immediately before harvesting samples, remove the tape backing.

2. Harvest the plant tissue and lay it on the double-sided tape with the surface to be imaged facing upward. Take care not to damage the tissue sample during handling.

3. Using a gentle stream of nitrogen, flatten any parts of the tissue that are not firmly attached to the double-sided tape.

4. Immediately start the sample drying step (Subheading 3.3) to minimize metabolite turnover.

3.3 Sample Drying

This procedure covers warming of cryosections and drying of tissue samples to quench metabolite turnover with minimal metabolite redistribution before the subsequent matrix application and MS imaging procedures.

1. Place the glass slides with the Cryo-Jane tape and tissue sections onto the cooled heat sink and immediately place into a

vacuum chamber. For directly attached samples, the heat sink is not necessary and samples can simply be dried under vacuum at room temperature.

2. Evacuate the chamber. Monitor the samples to ensure that no condensation occurs on the sample surface during the thaw-vacuum dry process (*see* **Note 13**).

3. After samples are dried and the heat sink is warmed sufficiently that water will not condense onto the sample when exposed to atmosphere (*see* **Note 2**), release the vacuum and remove the samples.

3.4 Matrix Application by Sublimation

This procedure allows for homogeneous application of organic matrices, especially those that do not give homogeneous coatings by traditional methods. For instance, 2,5-dihydroxybenzoic acid, the most commonly used matrix for MS imaging, is well known to form microcrystals of hundred micron size when spotted or sprayed. The sublimation procedure is modified from one described by Hankin and co-workers [15].

1. Attach the glass slide with the Cryo-Jane tape and tissue section to the bottom of the sublimation condenser using adhesive tape. If a smaller size sublimation apparatus is used, the Cryo-Jane tape with the attached tissue section should be transferred from the original glass slide to an appropriately sized glass slide first.

2. Evenly distribute ~300 mg of matrix over the bottom surface of the bottom flask of the sublimation apparatus and assemble the apparatus. The tissue section should be facing downward directly over the matrix on the bottom of the flask.

3. Evacuate the sublimation apparatus to <100 mtorr. Once vacuum is reached, add crushed dry ice and ~10–20 mL of acetone to the condenser reservoir to form a slurry and cool the tissue sample (*see* **Note 14**).

4. After 2–3 min of cooling, place the sublimation apparatus into the preheated mantle to initiate sublimation.

5. After the desired amount of matrix has been deposited, remove the sublimation apparatus from the heating mantle (*see* **Note 15**). Carefully half-fill the condenser reservoir with room-temperature water and wait for 2–3 min for the sample to return to room temperature.

6. Disassemble the sublimation apparatus and pour the water from the condenser reservoir, taking care to avoid splashing water onto the sample.

7. Remove the glass slide and sample from the bottom of the condenser, then remove the Cryo-Jane tape from the glass slide, and attach it to the MALDI sample plate using adhesive tape (*see* **Note 16**).

3.5 Matrix Application by Oscillating Capillary Nebulizer (OCN)

This procedure allows homogeneous application of inorganic matrices that cannot be sublimated.

1. Remove the Cryo-Jane tape with tissue sections from the glass slides and attach them to a MALDI sample plate using adhesive tape. Alternatively, the glass slide with tissue section can be directly used in some MALDI mass spectrometers (*see* **Note 1**).

2. Fill the syringe with the matrix solution/suspension.

3. Place a blank stainless steel plate 8–10 cm below the tip of the OCN. This will coat an area approximately 1–2 cm in diameter.

4. Adjust the nebulizing gas pressure to ~40 psi and start the gas flow.

5. Set the flow rate of the syringe pump at 50 μL/min.

6. Start the flow and monitor the blank stainless steel surface to ensure that matrix is being applied and wetting is minimal. Depending on matrix solution composition, it may be necessary to stop occasionally for complete drying (e.g., 10 s for every 5 s of spraying). Adjust the interval of application, flow rate, and stop time if necessary.

7. Stop the matrix flow and place the tissue slide below the tip of the OCN, and repeat **step 6** to apply the matrix to the tissue. For larger samples, the sample can be moved underneath the spray (*see* **Note 17**).

8. When a suitable amount of matrix is deposited (*see* **Note 5**), stop the syringe pump and remove the sample slide from below the OCN.

3.6 Multiplex MS Imaging

Please refer to the instrument guideline or manual for the operational details of MS imaging. Here, we describe only the basic procedure with a focus on multiplex MS imaging. The procedure described here is intended for LTQ-Orbitrap instruments, but the idea can be generalized to other mass spectrometers.

1. Using either an optical scanner or in-source camera, acquire an optical image of the whole tissue for later reference and co-registration with the MS images.

2. Using the instrument software (Tune Plus), optimize instrumental parameters for the studied tissue. This may include laser energy per pulse, number of laser shots, and ion optical tuning parameters (*see* **Note 18**). Save the tune file that contains this information.

3. Decide the type of multiplex MS imaging experiment that will be performed. Fig. 3 offers two examples that we have previously reported [10, 13]. The diagrams illustrate the spiral pattern on each raster step and the tables show the corresponding MS event for each spiral step, as defined in the method file

Raster Design

a

2 IT MS^2 m$_1$	3 IT MS^2 m$_2$
1 FT MS	4 IT MS^2 m$_3$

b

9 IT MS	2 IT MS^2 m$_1$	3 IT MS
8 IT MS^3 m$_2$→m$_4$	1 FT MS	4 IT MS^2 m$_2$
7 IT MS	6 IT MS^3 m$_1$→m$_3$	5 IT MS

MS Scan Event

Scan event #	Analyzer	MS^n Setting
1	FT	–
2	IT	MS^2 of m$_1$
3	IT	MS^2 of m$_2$
4	IT	MS^2 of m$_3$

Scan event #	Analyzer	MS^n Setting
1	FT	–
2	IT	MS^2 of m$_1$
3	IT	–
4	IT	MS^2 of m$_2$
5	IT	–
6	IT	MS^2 of m$_1$ MS^3 of m$_3$
7	IT	–
8	IT	MS^2 of m$_2$ MS^3 of m$_4$
9	IT	–

Fig. 3 Example spiral patterns for each raster step and corresponding MS or MS/MS events for two multiplex imaging experiments. Pattern (**a**) shows an experiment that acquires one high-mass-resolution orbitrap spectrum (step #1) and MS/MS spectra for three ions (steps #2–4) at each pixel. Pattern (**b**) shows an experiment that acquires one high-mass-resolution orbitrap spectrum (step #1), four moderate-mass-resolution but high-scan-speed ion trap MS spectrum (steps #3, 5, 7, and 9), MS/MS spectra for two ions (steps 2 and 4), and MS^3 spectra for two fragments of those ions (steps 6 and 8). FT and IT denote orbitrap and ion trap mass analyzer, respectively

(*see* **Note 19**). Polarity switching can also be integrated to acquire MS and MS/MS spectra in both positive and negative ion mode, as we have previously demonstrated [16]. It is also possible to incorporate data-dependent MS/MS or MS^n scans, with peaks for tandem MS analysis being chosen "on the fly" based on an MS spectrum for each raster pixel.

4. In the MALDI window of Tune Plus, select the tissue region to be imaged. Define the number of spiral steps, which should be the same as the number of MS scan events of the desired raster design, e.g., 4 and 9 for the raster design of A and B in Fig. 3, respectively. Define the raster step size and spiral step size. Make sure that the spiral step size is bigger than the laser spot size (*see* **Note 20**) and the raster step size is at least twice

the spiral step size for design A and four times the spiral step size for design B (*see* **Note 21**). Save all this information as a MALDI Position file.

5. Set up a new instrument method using the Xcalibur software. Individually define parameters for each scan event, such as the mass analyzer and MS^n settings, to match your raster design. Additional information is needed including mass range, desired resolution, polarity, and MS/MS conditions (*see* **Notes 22** and **23**). Save the instrument method file.

6. In the Sequence Set-up window of Xcalibur, provide a data file name to be used, destination folder for data, instrument method, and MALDI position file name.

7. (Optional) If multiple runs will be performed, repeat **steps 3–6**. **Step 3** can be skipped if the same multiplex imaging method will be used for the other imaging runs.

8. In the sequence window, select the samples to be analyzed and submit. Make sure that the submitted samples show up in the acquisition queue. Plate movement and laser firing should be visible on the in-source camera, and acquisition of spectra should be seen in the Tune window or the real-time view of Xcalibur.

---

4   Notes

1. MALDI-ion trap mass spectrometers (with or without Orbitrap) generally require a relatively low voltage to extract ions from the MALDI source (e.g., ±20 V) and accumulate a negligible surface charge during the MALDI process. In addition, subsequent ion trap or Orbitrap mass measurement is independent of initial kinetic energy and is not affected by minor electric field distortions on the MALDI plate surface. Accordingly, one can use non-conductive materials, including non-conductive adhesive tape and glass sample slides. For other instrumentation, especially time-of-flight (TOF) mass analyzers, this effect could be detrimental and one should use only conductive tape and plates.

2. The block should be large enough to keep attached tissues cold during the vacuum-drying process. We found that an ~200 g (approx. 18 cm × 6 cm × 0.65 cm) aluminum block is sufficient for 10–20 μm cryosections. Using this setup, the freeze-drying process takes ~90 min before samples are dried and sufficiently warmed to be removed from the vacuum.

3. The cold trap is submerged in a dry ice/acetone slurry and reduces contamination of the vacuum pump by matrix. It also minimizes backflow of vacuum pump oil to the sample.

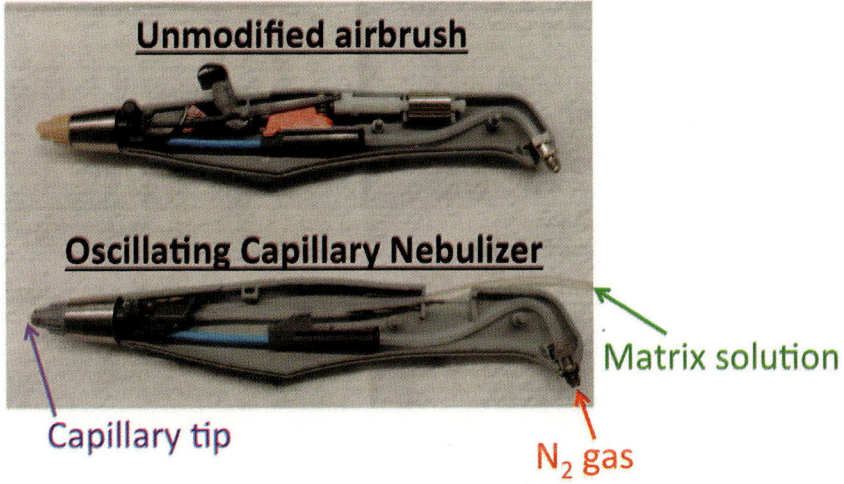

Fig. 4 Photograph showing modifications to an Aztek A470 airbrush (Testor, Rockford, IL) to create an oscillating capillary nebulizer. The trigger mechanism has been removed and a capillary has been run through the housing and out the spray tip. The capillary is run through plastic tubing (0.0625″ o.d. × 0.02″ i.d.) and nitrogen gas is supplied using the airbrush gas interface. Matrix solution is delivered through the capillary by means of a syringe pump

4. The choice of matrix is critical in MALDI-MS imaging experiments, especially for small metabolite analysis. Because of the wide chemical functionality of various metabolites, the matrix should be carefully chosen based on the analytes of interest. Matrix choice is discussed in [2].

5. Matrix concentrations and volumes should be optimized depending on the matrix and application area. We use 0.6 mL for 20 ppm colloidal silver suspension and 0.3 mL for colloidal graphite aerosol spray (Alfa Aesar, Ward Hill, MA), after dilution with 2-propanol four times and eight times, respectively.

6. The design and performance of an OCN are described in detail in Chen et al. [17]. A methodology for using an OCN for matrix application has also been published in this series [18]. We have created an OCN by modification of a commercial airbrush (Aztek A470; Testor, Rockford, IL). This modification is shown in Fig. 4. The inner spraying tip and trigger mechanism has been replaced by plastic tubing (0.02″ i.d. × 0.0625″ o.d.) with a fused silica capillary running through it (100 μm i.d. × 360 μm o.d.). Nitrogen is supplied as a nebulizing gas through the airbrush gas interface and matrix solution is supplied through the capillary by a syringe pump. The capillary can be easily replaced when switching matrix solutions to avoid cross-contamination.

7. Letting the gelatin completely freeze on liquid nitrogen can cause the tissue block to crack. Transferring to the cryostat right before the block is completely frozen can minimize this.

8. The gelatin should be opaque when completely frozen.

9. OCT can cause significant interference and suppression of analyte signals during MS analysis. Care should be taken to minimize the amount of OCT used to fix the tissue block, and to avoid squeezing the OCT over the sides of the sample.

10. Typical tissue sectioning for optical imaging uses thaw mounting by directly attaching a frozen tissue slice to a room-temperature glass slide. However, this may cause redistribution of water-soluble small molecules, which could be problematic for high-resolution MS imaging. We use adhesive Cryo-Jane tape followed by vacuum-drying to minimize this possibility.

11. We use the glass slides to make sections on the Cryo-Jane tape easier to transport and store, and as a heat sink to prevent thawing of the frozen tissue samples during transfer. Sample sections on Cryo-Jane tape thaw very quickly when exposed to ambient temperature.

12. Exposure of the samples to room-temperature air should be minimized. This both reduces the chances of sample thawing and prevents large amounts of ice condensing on the slide and sample.

13. Loss of some volatile analytes is unavoidable during freeze-drying, but these analytes are also likely to be lost in the vacuum or intermediate vacuum of most MALDI-MS ion sources. We do not recommend going below ~50–100 mtorr, so as not to lose partially volatile compounds that are amenable for analysis in an intermediate-pressure MALDI source. Vacuum drying should be performed even for atmospheric pressure MALDI-MS to quench metabolic turnover before matrix application and MS data acquisition.

14. Only a minimal amount of dry ice/acetone is needed. Only the bottom surface of the condenser (in contact with the sample) needs to be cooled. Using larger quantities will increase the time needed to rewarm the sample after sublimation.

15. After sublimation, a thin, even layer of matrix should be observable. If the matrix layer is completely opaque, the deposited layer is likely too thick. Thomas et al. found ~50–200 µg/cm² to be the optimal matrix density for analyte detection [19].

16. Ensure that the tape window with attached sample is flat against the MALDI plate. Variations in sample height will cause differences in laser fluence at the sample surface and possible signal deviation or spectral differences.

17. This method is best suited for spraying small tissue samples (<1 cm²). For larger samples, the motion of the tissue under the OCN should be automated to ensure an even coating, and the sprayed matrix volume should be adjusted accordingly.

18. These parameters are specific to the tissue sample, instrument, and matrix used, and it is very helpful to test and tune them on a section of dummy tissue that was processed in parallel. As a guideline, we generally use a laser energy of 1–10 μJ per pulse at 60 Hz, 10 shots per scan, and no sweep shots.

19. The "spiral step" function in Tune Plus is originally intended to average several spectra over a single pixel, incorporating several smaller steps into each raster step. We define the number of spiral steps and number of MS scans such that each spiral step correlates to a given MS scan type.

20. Laser spot size can be estimated by rastering over a thin matrix layer prepared by spotting α-cyano-4-hydroxycinnamic acid in acetone on a MALDI plate, with the raster size at least twice the expected laser spot size.

21. This pattern requires additional space equal to the size of one spiral raster step between each raster pattern. This preserves the spacing between ion trap MS scans and provides twice the spatial resolution of other scans in the pattern for imaging. Note that homemade software is required for generating images from IT scans, as ImageQuest does not process individual spiral steps as separate pixels.

22. For high-resolution analyzers, it is often desirable to collect data in a centroid or binned mode. Imaging of large tissues or at high spatial resolutions generates very-large-size data files (e.g., >1 GB) if full profile scans are collected.

23. For semi-targeted analysis, a list of parent m/z values can be used to acquire MS/MS or MS^n spectra for compounds known or suspected to be present in the tissue. Scans can also be acquired simply for the highest intensity peaks. It is helpful to incorporate a dynamic exclusion function, which prevents repeated acquisition of spectra for the most abundant ions and maximizes the number of ions for which tandem MS scans are performed. However, we have observed that if no peaks are selected for MS/MS in a data-dependent scan, the software will skip the scan without skipping the position. This can lead to acquisition of multiple Orbitrap scans within a single spiral pattern. Although this is generally not detrimental to imaging (multiple identical scans within a spiral pattern are simply averaged during image generation and this problem rarely occurs), it can increase data acquisition time.

Acknowledgments

We acknowledge kind advice from Dr. Zhihong Song, Dr. Basil Nikolau, Dr. Harry Horner, and Tracey Pepper in developing the sample preparation method. Corn seeds and valuable input were provided by Dr. Marna Yandau-Nelson. This work was supported by the US Department of Energy (DOE), Office of Basic Energy Sciences, Division of Chemical Sciences, Geosciences, and Biosciences. The Ames Laboratory is operated by Iowa State University under DOE Contract DE-AC02-07CH11358.

References

1. Kaspar S, Peukert M, Svatos A, Matros A, Mock H-P (2011) MALDI-imaging mass spectrometry – An emerging technique in plant biology. Proteomics 11:1840–1850

2. Lee YJ, Perdian DC, Song Z, Yeung ES, Nikolau BJ (2012) Use of mass spectrometry for imaging metabolites in plants. Plant J 70: 81–95

3. Fiehn O (2002) Metabolomics–the link between genotypes and phenotypes. Plant Mol Biol 48:155–171

4. Hanhineva K, Rogachev I, Kokko H, Mintz-Oron S, Venger I, Karenlampi S, Aharoni A (2008) Non-targeted analysis of spatial metabolite composition in strawberry (Fragaria x ananassa) flowers. Phytochemistry 69:2463–2481

5. Horn PJ, Korte AR, Neogi PB, Love E, Fuchs J, Strupat K, Borisjuk L, Shulaev V, Lee YJ, Chapman KD (2012) Spatial mapping of lipids at cellular resolution in embryos of cotton. Plant Cell 24:622–636

6. Cha S, Song Z, Nikolau BJ, Yeung ES (2009) Direct profiling and imaging of epicuticular waxes on Arabidopsis thaliana by laser desorption/ionization mass spectrometry using silver colloid as a matrix. Anal Chem 81:2991–3000

7. Jun JH, Song Z, Liu Z, Nikolau BJ, Yeung ES, Lee YJ (2010) High-spatial and high-mass resolution imaging of surface metabolites of Arabidopsis thaliana by laser desorption-ionization mass spectrometry using colloidal silver. Anal Chem 82:3255–3265

8. Burrell M, Earnshaw C, Clench M (2007) Imaging matrix assisted laser desorption ionization mass spectrometry: a technique to map plant metabolites within tissues at high spatial resolution. J Exp Bot 58:757–763

9. Lunsford KA, Peter GF, Yost RA (2011) Direct matrix-assisted laser desorption/ionization mass spectrometric imaging of cellulose and hemicellulose in Populus tissue. Anal Chem 83:6722–6730

10. Perdian DC, Lee YJ (2010) Imaging MS methodology for more chemical information in less data acquisition time utilizing a hybrid linear ion trap – orbitrap mass spectrometer. Anal Chem 82:9393–9400

11. Shrestha B, Patt JM, Vertes A (2011) In situ cell-by-cell imaging and analysis of small cell populations by mass spectrometry. Anal Chem 83:2947–2955

12. Li B, Bjarnholt N, Hansen SH, Janfelt C (2011) Characterization of barley leaf tissue using direct and indirect desorption electrospray ionization imaging mass spectrometry. J Mass Spectrom 46:1241–1246

13. Yagnik GB, Korte AR, Lee YJ (2013) Multiplex mass spectrometry imaging for latent fingerprints. J Mass Spectrom 48:100–104

14. Chaurand P, Cornett DS, Angel PM, Caprioli RM (2011) From whole-body sections down to cellular level, multiscale imaging of phospholipids by MALDI mass spectrometry. Mol Cell Proteomics 10(O110):004259

15. Hankin JA, Barkley RM, Murphy RC (2007) Sublimation as a method of matrix application for mass spectrometric imaging. J Am Soc Mass Spectrom 18:1646–1652

16. Korte AR, Lee YJ (2013) Multiplex mass spectrometric imaging with polarity switching for concurrent acquisition of positive and negative ion images. J Am Soc Mass Spectrom 24: 949–955

17. Chen Y, Allegood J, Liu Y, Wang E, Cachon-Gonzalez B, Cox TM, Merrill AH Jr, Sullards MC (2008) Imaging MALDI mass spectrometry using an oscillating capillary nebulizer matrix coating system and its application to analysis of lipids in brain from a mouse model of Tay-Sachs/Sandhoff disease. Anal Chem 80: 2780–2788

18. Chen Y, Liu Y, Allegood J, Wang E, Cachon-Gonzalez B, Cox TM, Merrill AH Jr, Sullards MC (2010) Imaging MALDI mass spectrometry of sphingolipids using an oscillating capillary nebulizer matrix application system. Methods Mol Biol 656:131–146

19. Thomas A, Charbonneau JL, Fournaise E, Chaurand P (2012) Sublimation of new matrix candidates for high spatial resolution imaging mass spectrometry of lipids: enhanced information in both positive and negative polarities after 1,5-diaminonaphthalene deposition. Anal Chem 84:2048–2054

Chapter 7

DESI Imaging of Small Molecules in Biological Tissues

Elaine C. Cabral and Demian R. Ifa

Abstract

Desorption electrospray ionization (DESI) allows the direct analysis of ordinary objects or preprocessed samples under ambient conditions. Among other applications, DESI is used to identify and to record spatial distributions of small molecules in situ, sliced or imprinted biological tissue. Manipulation of the chemistry accompanying ambient analysis ionization can be used to optimize chemical analysis, including molecular imprinting. Images are obtained by continuously moving the sample relative to the DESI sprayer and the inlet of the mass spectrometer. The acquisition time depends on the size of the surface to be analyzed and on the desired resolution.

Key words Mass spectrometry, Imaging, Biological tissues, Blotting, Small molecules

1 Introduction

Mass spectrometry imaging (MSI) has become an important technique in materials science and is an emerging area in the biological and forensic sciences [1]. The spatial distribution of chemical constituents of a sample and the information about the relative intensities of the ions allow the creation of detailed 2D images specific to particular chemicals. Secondary-ion mass spectrometry (SIMS) [2], matrix-assisted laser-desorption ionization (MALDI) [3], and desorption electrospray ionization (DESI) [4] are most commonly used and these techniques have been described for imaging of different kinds of samples [1, 5]. Among these techniques, DESI-MS imaging has become increasingly attractive, because of its simplicity and the reduced sample preparation steps compared with vacuum imaging techniques [6].

DESI is a family member of the ambient ionization techniques. In ambient ionization the surface is sampled with minimal or no preparation, ionization occurs externally to the mass spectrometer, and ions, not the entire sample, are introduced into the mass spectrometer [7]. In DESI, a spray of charged droplets generated by ESI is directed to the sample surface creating a thin film of solvent

Lin He (ed.), *Mass Spectrometry Imaging of Small Molecules*, Methods in Molecular Biology, vol. 1203,
DOI 10.1007/978-1-4939-1357-2_7, © Springer Science+Business Media New York 2015

on the surface. Further, spray droplets collide with the film splashing secondary droplets containing dissolved analytes into the air from which they are introduced into the mass spectrometer [8]. In addition, the ionization by DESI is soft, with an average internal energy deposition of ~2 eV in typical cases, which is similar to the internal energy of electrospray ionization (ESI) [9]. Hence DESI and other ambient imaging methods yield minimal fragmentation [10], and they are ideal for imaging in the low-mass region, without interference from added matrix (cf. MALDI) or fragmentations of large molecules (cf. SIMS). Identification of analytes can be achieved by using tandem mass spectrometry (MS/MS) or accurate mass measurements.

1.1 DESI Imaging Experimental Design

The workflow in MSI normally proceeds through three steps: (a) sample preparation, (b) data acquisition, and (c) data analysis (Fig. 1). Sample preparation is minimal in ambient ionization compared to traditional MSI. In some cases, such as imaging of

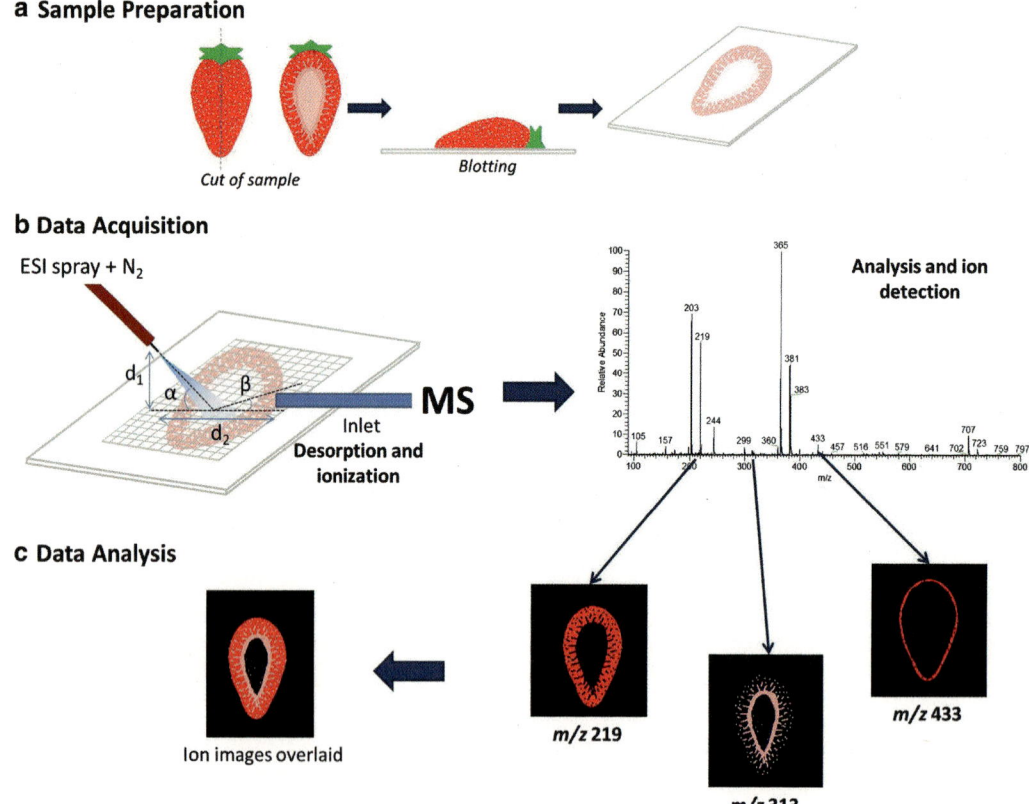

Fig. 1 MSI analysis workflow: (**a**) The first step is sample preparation, which requires minimal or sometimes no sample preparation; (**b**) data acquisition includes the steps of desorption/ionization, mass analysis, and ion detection, and (**c**) data analysis is the last step, where the spectra are converted in two-dimensional ion images using a color intensity scale with the relative ion intensity reflected by the intensity of the color

questionable documents or studies of natural products in plant material, the samples are ready for analysis without any sample preparation whatsoever. However, the geometry of the system does not allow direct imaging of soft or irregular surfaces such as whole animal and vegetable tissues [11]; this is because the success of imaging analysis by DESI is dependent on the angles and distances between the nozzle and the sample or the entry of MS. Changing these parameters during the data acquisition results in signal intensity changes [12]. Alternatively, the use of blotting or imprint techniques in which the chemicals are initially transferred to flat hard surfaces is an approach which has been successfully applied in MS imaging by MALDI [13–16], SIMS [17], nano-assisted laser-desorption ionization (NALDI) [18], and DESI [19–23]. Analysis of the spatial distribution of compounds in tissues by MS imaging is commonly performed using thin tissue sections obtained by cryosectioning bulk tissue in a cryostat [10]. For the study of animal tissues, sectioning and drying are still necessary. However, the sample is not introduced into the vacuum nor are chemicals added to the sample (matrices or tagging agents).

Data acquisition includes the steps of desorption/ionization, mass analysis, and ion detection. DESI-MSI data acquisition is performed by using microprobe mode as method for sample interrogation. In this mode, the whole surface of the sample is divided conceptually into small areas (pixels) which are scanned individually and sequentially in time. The steps of desorption and ionization in DESI are fully integrated and a single agent (the solvent) is employed. The ionization/desorption agent is set to provide the desired spatial resolution. The data from a single pixel are normally represented by a single mass spectrum or an average of two or more acquired mass spectra within the specific spot or by continuously rastering the surface with the ionizing agent. In the common microprobe mode, the use of an automated moving stage is required to assure the reproducibility of the scan velocity in order to accurately reproduce the geometry of the system [24, 25].

After the sample has been ionized, the ions either solvated in microdroplets or free gas-phase ions are directed into the mass spectrometer and mass analyzed. Several types of analyzers have been employed for ambient ionization MSI including linear triple quadrupoles (QqQ), quadrupole ion traps (Q-Trap), OrbiTraps, time of flight (ToF), and Fourier transform ion cyclotron resonance MS (FTMS). Mass analysis is followed by the data processing step, the third stage in the sequence of steps leading to creation of MS images. The recorded mass spectra are converted into a 2D image file, which can then be opened and visualized by imaging programs (e.g., the freeware BioMap or commercially available software, FlexImaging). Two-dimensional ion images are displayed using a color intensity scale with the relative ion intensity reflected by the intensity of the color. Appropriate contrast in the color bar and overlay can also be used to improve the visualization of the ion images. These ion images can be used to represent intensity distributions of ions of a single m/z ratio or they can be more complex representations, including ion populations that embody a particular chemical, biological, or bioinformatics parameter, for example, a principal component derived from a multivariate statistical analysis [26, 27].

1.2 Solvent, Substrate, and Geometrical Optimization

Optimization of the DESI sprayer is important for obtaining a high-quality MS image. The solvent system needs to be optimized to obtain an adequate signal level, depending on the sample analyzed and the target compounds [24]. The distances between the spray tip, the substrate, and the inlet of the mass spectrometer should be adjusted to obtain an appropriate small spray spot. The size and shape of the spray spot are also affected by the solvent being used, the flow rate, the DESI source tip dimensions, and the nebulizing gas pressure [26]. As indicated in studies of DESI sampling, there are three distinct regions in the sampling spot, with most desorption taking place in the inner region [28, 29]. It is important to have a well-defined spot on the surface to minimize redeposition and mixing of the sample during imaging.

Other important parameters are spatial resolution and sensitivity, which need to be balanced. The typical spatial resolution in DESI is ~200 μm, but this value can be reduced to 40 μm under particular operating conditions [30] even for biological tissue [31]. However, it should be kept in mind that increasing the spatial resolution will lead to a decrease of the signal intensity and hence affects the quality of ion images and the needed time to record them. The spray angle also plays an important role. Typically a spray angle of 52° (from the surface plane) provides higher signal intensity, whereas when the sprayer is perpendicular to the substrate (spray angle: 90°), the signal intensity decreases due to decreased secondary droplet transfer efficiency. This perpendicular spray, also called geometry-independent DESI, requires an enclosure and a coaxial return tube to optimize sensitivity [32].

1.3 Protocol Overview

This procedure is intended to illustrate the application of DESI-MSI of biological tissues with emphasis on small-molecule detection. Three different sample preparation procedures are presented: direct analysis of the sample surface, cryosectioning tissue sample [33], and imprinted sample surface [34]. The methodology for sample preparation should be chosen considering the sample characteristics. Samples that cannot be directly accommodated in front of the MS due to their irregular/soft surfaces or large sizes can be easily imprinted on an absorbent surface and then imaged without the need to use a cryostat. The efficiency of blotting is directly influenced by the chemical properties of the surfaces as well as the properties of the chemical compounds which are transferred into the surface.

2   Materials

2.1   Samples

1. Ginkgo leaves (*Gingko biloba* L.).
2. Zebra fish (*Brachydanio rerio*).
3. Strawberry fruits (*Fragaria × ananassa* Duch.).

2.2   Solvent and Reagents

1. Water (ultrapure, 18 MΩ-cm).
2. Methanol (MeOH: HPLC grade).
3. Acetonitrile (ACN: HPLC grade).
4. Compressed nitrogen (99.995 %).
5. Carboxymethyl cellulose.
6. Tricaine methanesulfonate (MS222).
7. Dry ice.

2.3   Molds, Surfaces, and Other Materials

1. Plain glass slides.
2. Plastic histological mailer.
3. TLC plates, 200 μm of thickness and 25 μm of pore size.
4. Aluminum foil.
5. Disposable embedding mold, rectangular 22 × 40 × 20 mm.
6. Embedding media (FSEM—Shandon Embedding Matrix).
7. Insulating container.
8. Desiccator.

2.4   Microtome and Mass Spectrometer

1. Cryomicrotome.
2. Microtome blades.
3. Tissue sample holder.
4. Hot plate or nonstick iron.
5. Syringe 500 μL.
6. Mass spectrometer (linear Ion-Trap LTQ Thermo Fisher Scientific, San Jose, CA—USA).
7. Extended ion-transfer tube (custom-built).
8. PicoTip emitter (New Objective, cat. no. TT150-50-50-N-5).
9. DESI source (custom-built or commercially available).

2.5   Software

1. Xcalibur 2.0 software (Thermo Fisher Scientific, San Jose, CA).
2. ImageCreator-v3.0 software.
3. BioMap (freeware, http://www.maldi-msi.org/).

3   Methods

3.1   Sample Preparation

3.1.1   Direct Analysis: Gingko Leaf

1. (Optional) Remove the cuticle leaf wax can by immersing the leaf in chloroform three times by 10 s each (*see* **Note 1**).
2. Attach the gingko leaf or a leaf piece on a flat surface (e.g., glass slide or metal plate) with a double-side tape (Fig. 2).

3.1.2   Histological Sections: Zebra Fish

1. First, prepare a tank with tricaine methanesulfonate solution (MS222, 200–300 mg/L) for fish euthanasia by prolonged immersion (*see* **Note 1**).
2. Remove the zebra fish from its main tank and place it in a tank with tricaine methanesulfonate solution. The fish should be left in the solution for at least 5–10 min following cessation of the opercular movement.
3. Remove the euthanized fish from water; let the excess water drain and place it in aluminum foil. The euthanized fish should be kept in the fridge until mold preparation (*see* **Note 2**).
4. Heat water (200 mL) in a beaker on a hot plate at 35–45 °C and add small portions (10 mg each time) of carboxymethyl cellulose (CMC) while stirring. The CMC should be added until the right level of viscosity, like a paste/gel consistency.
5. Fill the mold with CMC until 1/3 of total volume and put in the freezer by 25 min.
6. Remove the mold from the freezer and immediately place the zebra fish on top of the CMC, followed by the addition of more CMC to complete to mold (*see* **Note 3**).
7. Put the filled molds in the freezer at –8 °C overnight (Fig. 3).

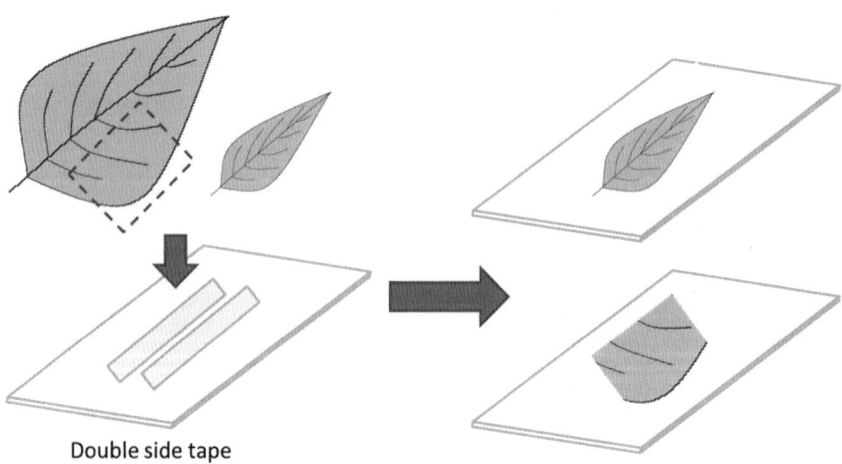

Double side tape

Fig. 2 Direct analysis preparation

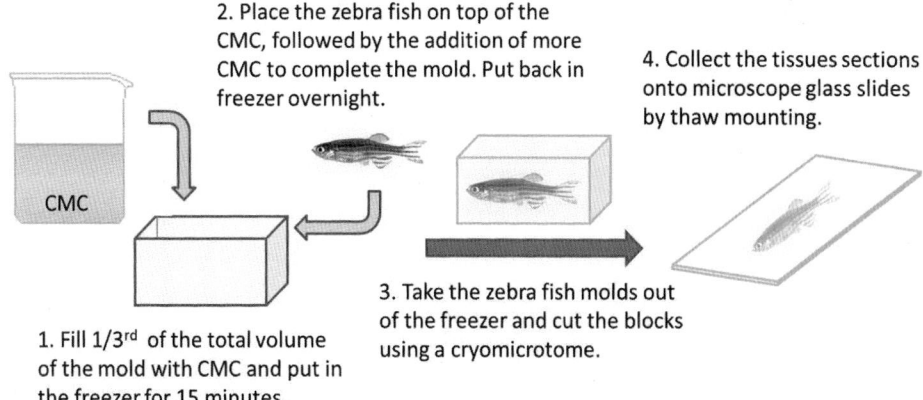

2. Place the zebra fish on top of the CMC, followed by the addition of more CMC to complete the mold. Put back in freezer overnight.

4. Collect the tissues sections onto microscope glass slides by thaw mounting.

CMC

3. Take the zebra fish molds out of the freezer and cut the blocks using a cryomicrotome.

1. Fill 1/3ʳᵈ of the total volume of the mold with CMC and put in the freezer for 15 minutes.

Fig. 3 Histological section preparation

8. Take the zebra fish molds out of the fridge and place it in an insulating container with dry ice to maintain subzero temperatures when transferring to the cryomicrotome station.

9. Remove the sample from the mold and place it on the sample holder using a minimal amount of frozen specimen-embedding media (FSEM). The use of FSEM should be minimized and restricted to the tissue attachment to the sample holder (*see* **Note 4**).

10. Cut the tissue into 20–45 μm thick slices using a cryomicrotome at –10 to –18 °C.

11. Collect the tissue sections onto microscope glass slides by thaw mounting. This is accomplished by attaching the cold tissue sections to the glass slide (Fig. 3).

12. (Optional) If the samples will not be immediately analyzed, save the slides in a closed container (e.g., plastic histological mailers) on dry ice and store at –80 °C until use.

13. (Optional) If the slides have been stored at –80 °C, take the closed containers with the slides from the freezer and let them warm up at room temperature (~21 °C) in a vacuum desiccator before analysis.

3.1.3 Blotting: Strawberry or Gingko Leaf

Direct Blotting

1. Cut the TLC plate in square shape with 5 cm × 5 cm or according to the sample size. Always use 2 cm more than the original sample size.

2. Manually cut the fresh strawberry in side view by using a common blade (e.g., kitchen knife).

3. Blot the TLC plates by positioning the sectioned strawberry directly onto the TLC plate for 5 s (Fig. 4).

4. For leaf imprinting put the leaf between two TLC plates and press against a hard surface (e.g., iron at room temperature) by 15 s.

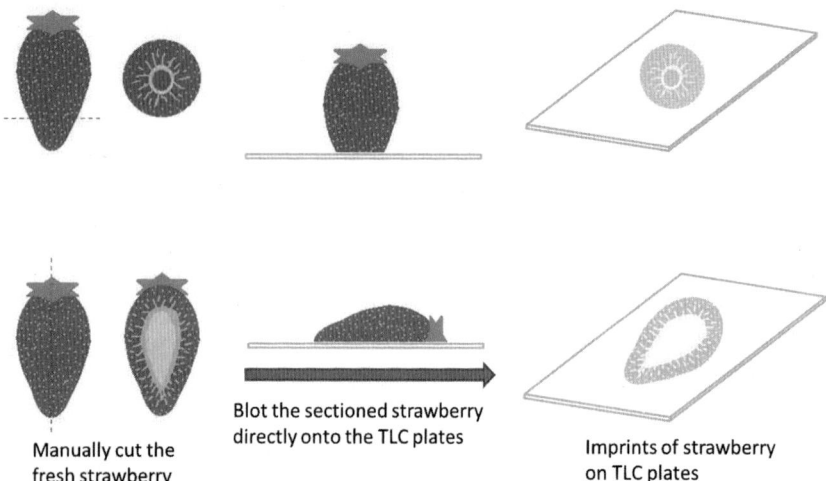

Fig. 4 Direct blotting preparation

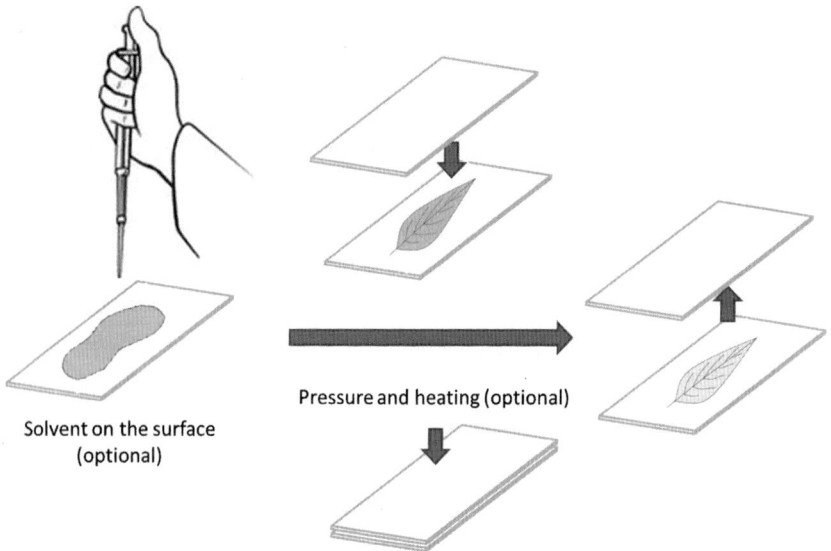

Fig. 5 Blotting assisted by heating and/or solvent extraction preparation

Thermo-Assisted Blotting

1. Prepare the TLC plate and cut the strawberry as described in Subheading 3.1.3.1 (direct blotting).

2. Heat the TLC plate in a hot plate or iron until 200 °C.

3. Blot the strawberry by positioning the sectioned strawberry directly onto the TLC plate for 5 s.

4. For leaf imprinting place the leaf between two TLC plates and press it against a hard and hot surface (e.g., iron at 200 °C) for 15 s (Fig. 5).

Solvent Extraction
Assisted Blotting

1. Prepare the TLC plate and cut the strawberry as described in Subheading 3.1.3.1 (direct blotting).

2. Wet the TLC plate with 0.5 mL of solvent using a pipette prior to making the imprint.

3. Blot the strawberry by positioning the sectioned strawberry directly onto the TLC plate for 5 s.

4. For the leaf imprinting, put the leaf between two wetted TLC plates with 0.5 mL of solvent and press the sandwich against a hard surface (e.g., iron at room temperature) for 15 s (Fig. 5).

Thermal Imprinting
with Solvent Extraction-
Assisted Blotting

1. Prepare TLC plate and strawberry cut in the same way as in the previous Subheading 3.1.3.1 (direct blotting).

2. Heat the TLC plate in a hot plate or iron until 200 °C.

3. Wet the TLC plate with 0.5 mL of solvent using a pipette prior to making the imprint.

4. Blot on TLC plates by positioning the sectioned strawberry directly onto the TLC plate for 5 s.

5. For the leaf imprinting put the leaf between two wetted TLC plates with 0.5 mL of solvent and press against a hard and hot surface (e.g., iron at 200 °C) for 15 s (Fig. 5).

3.2 Data Acquisition

1. Optimize the DESI system (ionization mode, solvent, flow rate, and geometry—Fig. 6) to obtain high-quality images. For the example shown here, the DESI parameters are listed in Table 1 (*see* **Notes 5–7**).

2. In order to obtain the correct setup for the required image resolution, divide the tissue sample area by the required pixel size. For instance, imprints of gingko leaves with dimensions of 33×20 mm^2 will result in a matrix of 165×100 pixels, each pixel covering an area of 200×200 μm^2 (*see* **Notes 8** and **9**).

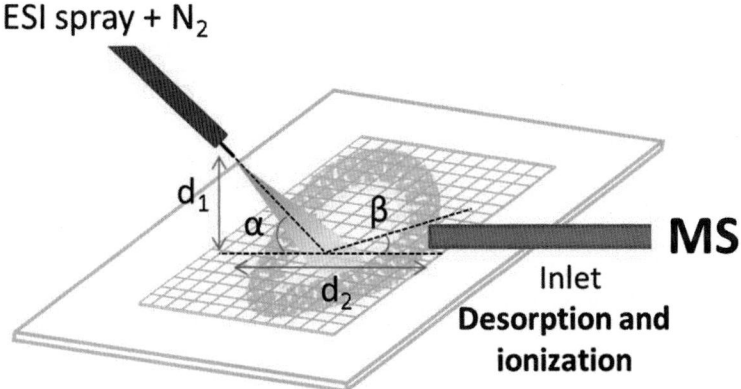

ESI spray + N$_2$

MS

Inlet
**Desorption and
ionization**

Fig. 6 Scheme of DESI source with important geometric parameters, d_1 = sprayer-to-surface distance, d_2 = sprayer-to-inlet distance, α = incident spray angle, and β = collection angle

Table 1
DESI parameters for Gingko leaf samples

Parameters	Strawberry
Solvent	MeOH (100 %)
Flow rate (μL/min)	3.0
Ionization mode	Negative
m/z range	120–1,200
Collection angle—β	10°
Incident spray angle—α	50°
Sprayer-to-inlet distance—d_2 (mm)	4.0
Sprayer-to-surface distance—d_1 (mm)	1.0

3. Calculate the acquisition time for each line scan based on the number of pixels and the scan time of one mass spectrum. For instance, 165 pixels × 0.48 s = 79.2 s. Calculate the travel velocity for the XY moving stage. In this example, 33 mm/ 79.2 s = 416 μm/s.

4. Create a sample list in the mass spectrometer acquisition software (Xcalibur 2.0). The total number of samples in the list is equal to the number of lines in the image. The last two characters of the file name should index the sequence of files, for instance, Gingko_01.raw, Ginkgo_02.raw … Gingko_100.raw. Make sure that the acquisition method contains the correct acquisition time for each line, for instance, 100 lines with an acquisition time of 79.2 s (1.32 min) each.

5. Set up the traveling velocity of the moving stage, the sample dimensions, and the pixel size on the XY moving-stage software controller.

6. Check if the syringe contains enough solvent to acquire the whole image. Let the spray stabilize for 1–2 min before starting. This is accomplished by turning the syringe pump, the nebulizing gas, and the high voltage on and waiting for 1–2 min, which is enough to fill the capillary with solvent under typical conditions and remove all air bubbles.

7. Start the acquisition.

3.3 Processing

1. Before data analysis, convert the Xcalibur 2.0 mass spectra files (.RAW extensions) into Analyze 7.5 format files (.img, .hdr and .t2m) required by BioMap. The software used was Img Converter v3.0 (*see* **Notes 10–12**).

2. Open the ImgConverter v3.0 software and follow the instructions on the main window: (a) fill out the name of the files to

be created; (b) fill out the number of pixels on each dimension (X and Y); (c) on the raw data box, find the first ".RAW" file acquired in the sample list. For example Gingko_01.raw; (d) click on the Append button. It should take only few minutes to create the files; (e) after all files are processed, the Exit button will become available; (f) click on Exit to close the program.

3. The following instructions illustrate how to generate chemical images of quinic acid (m/z 191) and bilobetin (m/z 551) using BioMap. Further information about the software, including tutorials about browsing, co-registration, and analysis, can be found at http://www.maldi-msi.org/. (a) Open the BioMap software; (b) click on the menu bar File>Import>MSI to load the converted file (from **step 2** above); (c) click on the menu bar Analysis>Plot>point to open the mass spectrum window; (d) on the mass spectrum window, expand the m/z range 150–600 by selecting this range while holding the right button of the mouse; (e) click with the left button on a peak. Click on the black and white scale bar on the top left-sided tool bar to change the color template. Select the rainbow color scale. Adjust the contrast of the image by selecting minimum and maximum values on the slide bars; (f) on the mass spectrum window click on the peak of m/z 191 (quinic acid, $[M-H]^-$). Adjust the contrast for better visualization. At this point, an image of the anthocyanidin distribution, such as Fig. 7a, should become visible; (g) copy the image by clicking on the menu bar Edit>Copy; (h) select a second window by clicking on it and paste the copied scan (Edit>Paste>Scan); (i) repeat steps (c) through (e) for the second window, now focusing on the m/z range 500–600. Select the peak at m/z 551. The distribution of the sugar should be visualized as in Fig. 7b.

4. The following instructions illustrate how to overlay the two chemical images obtained in **step 3** and display them simultaneously: (a) Take note of the "N" value, displayed at the bottom part of the spectrum window or at the central part of the left-sided toolbar, for instance, Position (191.369, 76.7340), #1004. This value corresponds to the ion m/z 191 on the table of mass (.t2m file); (b) in order to overlay the images, copy the second window scan (Edit>Copy). Select the first window and paste the copied data as overlay (Edit>Paste> As Ovl); (c) change the proprieties of the image by clicking the right button. Change the Overlay display mode (Ovldisplay modes) to Bicolor; (d) select the ion to be overlaid by clicking on the menu bar Window>Ovl Control; (e) set the overlay to m/z 219 by selecting the N value #1004 in the appropriated slide bar; (f) to improve contrast in this window, adjust the values of minimum and maximum. At this point, a plot, as shown in Fig. 7c, should be observed; (g) finally, present the

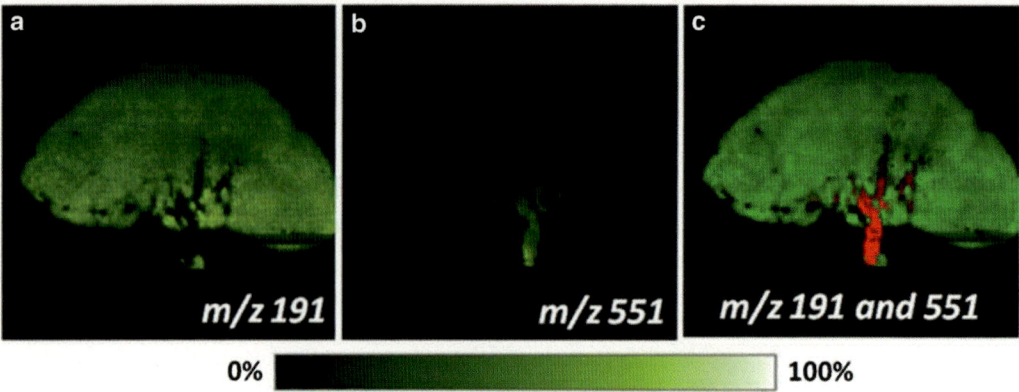

Fig. 7 Imaging analysis by DESI-MS of gingko leaves imprinted directly on TLC plates with solvent and heat assistance. Ion images show the spatial distribution of (**a**) ions *m/z* 191 (quinic acid); (**b**) ions of *m/z* 551 (bilobetin and (**c**) overlaid images. All ion images have same size (3.3 × 2.0 cm) and are plotted using the same color scale, which is depicted next to the panels to visualize the relative ion intensities from 0 % (*black*) to 100 % (*white*)

image as interpolation (default) or as voxels. Click on the image with the left mouse button, select properties, and change the display method to voxels. An image, as shown in Fig. 7c, should be observed.

4   Notes

1. CAUTION: Animals should be handled according to local Institutional Animal Care and Use Committee and Institutional Biosafety Committee-approved protocols.

2. The molds should be prepared the day before the cryosectioning.

3. CRITICAL STEP: Pay attention on the fish position inside the mold; it is determinant for a good sagittal cut in cryomicrotome. The fish position should be as flat as possible.

4. Be sure that the FSEM does not make any contact with the area of interest of the tissue. If this precaution is not taken, the components of the FSEM will dominate the mass spectra during data acquisition.

5. The Materials section, spare tissue material not intended for imaging can be used for the final adjustments to the DESI parameters, in particular, the spray solvent. The solvent composition can influence the ability to detect particular analytes from the tissue matrix depending on the solubility of the analyte in the solvent system.

6. Use silica capillaries with blunt cut ends for well-defined symmetrical elliptical spray spots. Commercially available tapered silica capillaries (e.g., PicoTip emitter tip: 50 ± 3 μm) have shown good results in our experiments. Spray optimization can be

carried out using water-sensitive paper from TeeJet to visualize the spray pattern on the surface as a function of operating parameters.

7. Ensure that the tip of the fused silica capillary is blunt and not burred or cracked. A burred or cracked tip will result in spray instability leading to irreproducible data.

8. The desorption/ionization agent(s) used will dictate the lateral and the depth resolution of the images acquired. The spatial resolution depends on the spot size of the ionizing agent but it can also be limited by other factors such as operating conditions (substrate moving speed, step size, analyte carryover between pixels, etc.).

9. Figure 6 illustrates the geometry parameters. The geometry and the flow rate are determined on the spot size and can be used as a guide for establishing optimal spray conditions. Therefore, if high resolution is required for a particular application it is recommended to use low solvent flow rates. An optimal spray pattern is obtained when well-defined elliptical spots are created with minimal forward "splashing" and negligible backward "splashing." * CRITICAL STEP: High solvent flow rates could result in excessive wetting of the surface and cross contamination.

10. Processing the imaging data, such as background subtraction or normalization, can be done on the mass spectra prior to image creation or directly on the chemical images.

11. The analyzer employed will dictate the mass resolution represented in the final data. When using a high-mass-resolution mass spectrometer for imaging, the data file size of the image can easily reach gigabytes (GB) in contrast to a few megabytes (MB) stored in low-resolution experiments of the same sized sample.

12. The data file size of the image is related to the number of pixels, the mass range used, and the mass resolution chosen. Processing and visualization of large-data-size ion images require computers with large memories and relatively high speed. By carefully setting these parameters, the data file size of the generated image can be made to fall in a range suitable for processing and storage, but the development of imaging software and hardware is an important ongoing task which is necessary to allow the full potential of high-mass-resolution imaging to be reached.

Acknowledgments

We thank the Brazilian National Council for Scientific and Technological Development (CNPq) and Natural Science and Engineering Research Council of Canada (NSERC).

References

1. McDonnell LA, Heeren RM (2007) Imaging mass spectrometry. Mass Spectrom Rev 26(4): 606–643

2. Belu AM, Graham DJ, Castner DG (2003) Time-of-flight secondary ion mass spectrometry: techniques and applications for the characterization of biomaterial surfaces. Biomaterials 24(21):3635–3653

3. Sinha TK, Khatib-Shahidi S, Yankeelov TE et al (2008) Integrating spatially resolved three-dimensional MALDI IMS with in vivo magnetic resonance imaging. Nat Methods 5(1):57–59

4. Ifa DR, Wu CP, Ouyang Z, Cooks RG (2010) Desorption electrospray ionization and other ambient ionization methods: current progress and preview. Analyst 135(4):669–681

5. Dill AL, Eberlin LS, Ifa DR, Cooks RG (2011) Perspectives in imaging using mass spectrometry. Chem Commun 47(10):2741–2746

6. Wu C, Dill AL, Eberlin LS et al (2012) Mass spectrometry imaging under ambient conditions. Mass Spectrom Rev 32(3):218–243

7. Cooks RG, Ouyang Z, Takats Z, Wiseman JM (2006) Ambient mass spectrometry. Science 311(5767):1566–1570

8. Costa AB, Cooks RG (2007) Simulation of atmospheric transport and droplet-thin film collisions in desorption electrospray ionization. Chem Commun 38:3915–3917

9. Nefliu M, Smith JN, Venter A, Cooks RG (2008) Internal energy distributions in desorption electrospray ionization (DESI). J Am Soc Mass Spectrom 19(3):420–427

10. Wu C, Ifa DR, Manicke NE, Cooks RG (2010) Molecular imaging of adrenal gland by desorption electrospray ionization mass spectrometry. Analyst 135:28–32

11. Ifa DR, Srimany A, Eberlin LS, Naik HR et al (2011) Tissue imprint imaging by desorption electrospray ionization mass spectrometry. Anal Methods 3(8):1910–1912

12. Kertesz V, Van Berkel GJ (2008) Scanning and surface alignment considerations in chemical imaging with desorption electrospray mass spectrometry. Anal Chem 80(4):1027–1032

13. Valdes-Gonzalez T, Goto-Inoue N, Hirano W et al (2011) New approach for glyco- and lipidomics - Molecular scanning of human brain gangliosides by TLC-Blot and MALDI-QIT-TOF MS. J Neurochem 116(5):678–683

14. Mullen AK, Clench MR, Crosland S, Sharples KR (2005) Determination of agrochemical compounds in soya plants by imaging matrix-assisted laser desorption/ionisation mass spectrometry. Rapid Commun Mass Spectrom 19(18):2507–2516

15. Bunch J, Clench MR, Richards DS (2004) Determination of pharmaceutical compounds in skin by imaging matrix-assisted laser desorption/ionisation mass spectrometry. Rapid Commun Mass Spectrom 18(24):3051–3060

16. Caprioli RM, Farmer TB, Gile J (1997) Molecular imaging of biological samples: localization of peptides and proteins using MALDI-TOF MS. Anal Chem 69(23):4751–4760

17. Sjövall P, Lausmaa J, Nygren H et al (2003) Imaging of membrane lipids in single cells by imprint-imaging time-of-flight secondary ion mass spectrometry. Anal Chem 75(14): 3429–3434

18. Vidová V, Novák P, Strohalm M et al (2010) Laser desorption-ionization of lipid transfers: tissue mass spectrometry imaging without MALDI matrix. Anal Chem 82(12):4994–4997

19. Eberlin LS, Dill AL, Golby AJ, Ligon KL, Wiseman JM, Cooks RG, Agar NYR (2010) Discrimination of human astrocytoma subtypes by lipid analysis using desorption electrospray ionization imaging mass spectrometry. Angew Chem Int Ed 49(34):5953–5956

20. Eberlin LS, Dill AL, Costa AB et al (2010) Cholesterol sulfate imaging in human prostate cancer tissue by desorption electrospray ionization mass spectrometry. Anal Chem 82(9): 3430–3434

21. Girod M, Shi Y, Cheng JX, Cooks RG (2010) Desorption electrospray ionization imaging mass spectrometry of lipids in rat spinal cord. J Am Soc Mass Spectrom 21(7):1177–1189

22. Lane AL, Nyadong L, Galhena AS et al (2009) Desorption electrospray ionization mass spectrometry reveals surface-mediated antifungal chemical defense of a tropical seaweed. Proc Natl Acad Sci U S A 106(18):7314–7319

23. Watrous J, Hendricks N, Meehan M, Dorrestein PC (2010) Capturing bacterial metabolic exchange using thin film desorption electrospray ionization-imaging mass spectrometry. Anal Chem 82(5):1598–1600

24. Wu CP, Ifa DR, Manicke NE, Cooks RG (2009) Rapid, direct analysis of cholesterol by charge labeling in reactive desorption electrospray ionization. Anal Chem 81(18):7618–7624

25. Tawiah A, Bland C, Campbell D, Cooks RG (2010) Solvent effects and the role of solubility in desorption electrospray Ionization. J Am Soc Mass Spectrom 21:572–579

26. Ifa DR, Wiseman JM, Song QY, Cooks RG (2007) Development of capabilities for imaging

mass spectrometry under ambient conditions with desorption electrospray ionization (DESI). Int J Mass Spectrom 259(1–3):8–15

27. Wiseman JM, Ifa DR, Venter A, Cooks RG (2008) Ambient molecular imaging by desorption electrospray ionization mass spectrometry. Nat Protoc 3(3):517–524

28. Green FM, Stokes P, Hopley C et al (2009) Developing repeatable measurements for reliable analysis of molecules at surfaces using desorption electrospray ionization. Anal Chem 81(6):2286–2293

29. Pasilis SP, Kertesz V, Van Berkel GJ (2007) Surface scanning analysis of planar arrays of analytes with desorption electrospray ionization-mass spectrometry. Anal Chem 79(15): 5956–5962

30. Kertesz V, Van Berkel GJ (2008) Improved imaging resolution in desorption electrospray ionization mass spectrometry. Rapid Commun Mass Spectrom 22(17):2639–2644

31. Campbell DI, Ferreira CR, Eberlin LS, Cooks RG (2012) Improved spatial resolution in the imaging of biological tissue using desorption electrospray ionization. Anal Bioanal Chem 404(2):389–398

32. Venter A, Cooks RG (2007) Desorption electrospray ionization in a small pressure-tight enclosure. Anal Chem 79(16):6398–6403

33. Chramow A, Hamid TS, Eberlin LS et al (2014) Imaging of whole zebra fish (Danio rerio) by desorption electrospray ionization mass spectrometry. Rapid Commun Mass Spectrom 28(19):2084–2088

34. Cabral EC, Mirabelli MF, Perez CJ, Ifa DR (2013) Blotting assisted by heating and solvent extraction for DESI-MS imaging. J Am Soc Mass Spectrom 24(6):956–965

Chapter 8

Desorption Electrospray Ionization Imaging of Small Organics on Mineral Surfaces

Rachel V. Bennett and Facundo M. Fernández

Abstract

Desorption electrospray ionization (DESI)-mass spectrometry facilitates the ambient chemical analysis of a variety of surfaces. Here we describe the protocol for using DESI imaging to measure the distributions of small prebiotically relevant molecules on granite surfaces. Granites that contain a variety of juxtaposed mineral species were reacted with formamide in order to study the role of local mineral environment on the production of purines and pyrimidines. The mass spectrometry imaging (MSI) methods described here can also be applied to the surface analysis of rock samples involved in other applications such as petroleum or environmental chemistries.

Key words Desorption electrospray ionization, Ambient mass spectrometry, Imaging, Prebiotic chemistry, Minerals, Formamide

1 Introduction

Desorption electrospray ionization (DESI) is a spray-based, ambient ionization technique that allows for the direct analysis of surfaces [1]. The working model for desorption and subsequent ionization by DESI is known as the "droplet pickup" model [2–4]. The first step in the analysis process begins with the formation of charged primary droplets by the DESI probe, which are propelled towards the surface with a high-pressure gas. The collision of these droplets with the surface results in the wetting of the surface, creating a thin film in which the analyte is dissolved by a solid–liquid microextraction mechanism [2]. Subsequent droplet collisions break the solvent layer and create many secondary droplets thrust towards the instrument inlet containing the analytes present on the surface [3, 4].

Chemical imaging by mass spectrometry is a powerful method by which to map spatial distributions of molecules to better understand their function in the system of interest. When performed under atmospheric conditions, using DESI or laser ablation-electrospray ionization (LAESI) [5], for example, sample pretreatment is simplified

Lin He (ed.), *Mass Spectrometry Imaging of Small Molecules*, Methods in Molecular Biology, vol. 1203,
DOI 10.1007/978-1-4939-1357-2_8, © Springer Science+Business Media New York 2015

while still maintaining the high quality of information obtained. In DESI imaging, the sample is scanned unidirectionally with line stepping using a software-controlled stage underneath the ionization probe. Through the time domain, m/z information is correlated with the chemical species' spatial distribution. Ambient mass spectrometry imaging techniques are also more conducive for large or irregularly shaped samples compared to traditional methods carried out under high vacuum.

Herein, we describe a procedure by which to image a granite surface involved in a chemical reaction, using the example of the thermal decomposition of formamide in the presence of a granite sample which yields a variety of compounds of biological interest [6]. Formamide is the first hydrolysis product of HCN [7], and is a plausible prebiotic starting material for the synthesis of nucleobases and their derivatives due to its stability and low volatility compared to water [8, 7]. While purine and other nitrogen heterocycles can be formed upon heating or irradiating neat formamide [9–11], previous studies have shown that the addition of inorganic catalysis, such as minerals, can significantly alter the class and yield of products from the reaction [12, 13]. The presence of juxtaposed mineral species on the granite surface allows for the examination of the effects of local mineral environment on the product mixture by direct imaging.

2   Materials

2.1   Formamide Reaction

1. Granite sample: Any granite sample with a desirable mineral composition and a smooth surface is suitable for imaging (*see* **Note 1**). The granites used in this experiment had pore depths not exceeding 100 μm as measured by optical profilometry (Wyco Noncontact Profilometer NT 3300, Veeco).

2. Formamide: 50 mL per reaction. ACS reagent grade, >99.5 %. Vessel: 100 mL glass beaker, aluminum foil to cover beaker during reaction.

3. Oven: capable of reaching 165 °C.

4. Adenine standard solution.

2.2   DESI Imaging

1. Laboratory tissue (e.g., Kimwipe).

2. Solvents: Methanol (LC-MS Ultra Chromasolv grade), water (ultrapure, 18 MΩ cm). For DESI experiments, a 7:3 mixture of methanol–water should be used at a flow rate of 3 μL/mL.

3. Gas: Nitrogen, 140 psi.

4. Source geometry optimization sample: Omni Slide™ (Prosolia Inc.) or glass microscope slide, red (ultra) fine point Sharpie®.

5. DESI source: Omni Spray 2D automated DESI ion source (Prosolia Inc.).

6. Mass spectrometer: Synapt G2 HDMS mass spectrometer (Waters) operated in positive ion mode. MS acquisition parameters are input through the MassLynx software (Waters).

7. Image processing: Mass spectral data files are converted into the Analyze 7.5 data cube format using the FireFly™ (Prosolia Inc.) software for visualization in BioMap (maldi-msi.org).

3   Methods

3.1   DESI Source Geometry Optimization

The source geometry should be optimized prior to each experiment for maximum sensitivity.

1. Turn on the mass spectrometer.

2. For an approximate optimization of the source geometry, mark a shape ~3 mm² on the Omni Slide™ or microscope slide with the red Sharpie®.

3. Set the nebulizing gas and solvent flow rates to the values given in Subheading 2.2.

4. Position the DESI source so that the distance between the spray tip and the sample surface should be 2 mm, at an angle of 58° to the surface, and the separation between the source sprayer and the capillary inlet of the mass spectrometer set at 4 mm (*see* **Note 2**).

5. Allow ~20 min for the DESI spray to stabilize before continuing with the analysis. Optimize the source-surface-transfer capillary geometry (spray tip-sample surface distance, spray tip-inlet distance, inlet-sample surface distance, source angle, etc.) for maximum signal for *m/z* 443 (Rhodamine 6G dye present in the red Sharpie®).

3.2   Granite Surface Leveling

In order to successfully image the granite surface, the sample must be mounted so that the surface is perfectly level. Any variation in sample height due to tilt will alter the source-surface-inlet geometry and subsequently affect the data quality [15].

1. Mount the granite sample on a glass microscope slide using double-sided tape or modeling clay. If the sample does not have two smooth and parallel sides and more closely resembles a wedge shape, use cushioned tape or the modeling clay to support the granite sample in such a way that the top surface to be imaged is leveled.

2. In order to determine if the sample surface is level, place the sample in the sample holder of the stage and adjust the sample height so that the transfer capillary is resting just above (<1 mm, but not touching) the sample surface in the top left corner.

3. Move the sample underneath the capillary from the left to right and top to bottom. Any changes in spacing between the transfer capillary and the sample should be corrected for by removing or adding additional layers of tape or adjusting the modeling clay until no changes in capillary height are observed during repeated motions across the granite surface (*see* **Note 3**).

4. With the granite substrate in place, adjust the DESI source position as necessary for maximum sensitivity. An approximate optimal geometry is given in Subheading 3.1 (#4) and was further optimized in Subheading 3.1. With the source on, the geometry is considered optimal when the highest signal-to-noise ratio is achieved for solvent contaminants or expected product ions present on the surface.

3.3 Imaging Data Acquisition

1. Follow instructions in the Omni Spray User Manual to set up automated mass spectral data acquisition.

2. Set the scan speed to 200 μm/s and a line step height of 200 μm resulting in a pixel size of 200 μm^2 (*see* **Notes 4** and **5**).

3. In MassLynx, set up a Sample List with the same number of samples as image rows (in this case 50) giving each sample an appropriate and unique name (i.e., SampleA_01, SampleA_02, etc.).

4. Set the acquisition time for each sample to the time given by the Omni Spray software (50 s) for a *m/z* range of 50–300. Leave all other Sample List options (i.e., injection volume) that pertain to liquid chromatography experiments blank.

5. Begin the mass spectrometer acquisition by clicking "Run."

6. The mass spectrometer will then go through a series of initialization steps. When the status bar says "Waiting on Inlet 1," click "Start" in the Omni Spray software. This will begin the automated image data acquisition, and no operator input is needed until the image has completed.

3.4 Imaging Data Processing

1. Follow procedures outlined in the FireFly™ user manual for data conversion and processing.

3.5 Pre-reaction Cleaning and Blank Image

1. Clean the granite sample by heating in the oven at 165 °C for 24 h in order to remove any intrinsic organic material (*see* **Note 6**).

2. Acquire a DESI image of the unreacted surface using the high-pressure spray of DESI to further clean the granite by following steps outlined in Subheadings 3.1–3.4.

3. The image acquired of the unreacted granite surface is shown in Fig. 1.

3.6 Reaction Conditions

1. Place the granite sample in the 100 mL glass beaker or other vessel of choice.

2. Add 50 mL of formamide to the beaker (*see* **Note 7**).

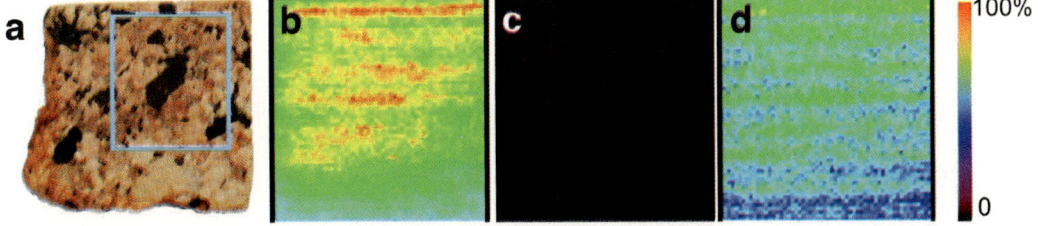

Fig. 1 (**a**) Optical image of the unreacted granite sample with the area imaged outlined in *blue*, (**b**) total ion image and selected ion images of (**c**) protonated purine (*m/z* 121) and (**d**) solvent contaminant phthalate ester (*m/z* 149) obtained by imaging the unreacted granite sample by DESI-MSI. The intensity scale ranges for (**c**) and (**d**) were set to match Fig. 1e, j, respectively, in order to demonstrate relative ion abundance difference before and after the formamide reaction. Reproduced from [6] with permission from American Chemical Society

3. Cover the beaker with aluminum foil.

4. Place the covered beaker in the oven (already heated to 160 °C) and ensure that the oven will not be disturbed (door opened, etc.) during the reaction time.

5. Leave the beaker in the oven for 96 h.

6. At the end of the reaction period, remove the sample from the oven and allow it to cool to room temperature.

7. Remove the granite sample from the reaction supernatant and blot with a laboratory tissue.

8. Store the reaction supernatant for later analysis if interested.

9. Immediately prepare the sample for imaging; otherwise store the granite in a Petri dish (or other container) in a refrigerator until ready for analysis.

3.7 Rinsing, Soaking, and Imaging

The supernatant of this reaction is viscous and the reaction products of interest were those closest to the granite surface. Therefore we carried out a series of rinsing steps in order to directly analyze the surface (*see* **Notes 8** and **9**).

1. Rinse the granite sample by pouring or pipetting 1 mL of the methanol–water mixture (7:3) over the surface and repeat for a total of two rinses.

2. Mount the granite sample and prepare for DESI analysis and imaging as outlined in Subheadings 3.1 and 3.2.

3. Image the granite surface as described in Subheadings 3.1–3.4.

4. If there are known products of interest, plot the ion's spatial distribution in BioMap using the FireFly-converted data.

5. Reaction products tentatively identified in this example reaction are listed in Table 1, and their spatial distributions are mapped in Fig. 2.

Table 1 Tentatively identified products and their distributions across minerals present in the granite sample shown in Fig. 2

Assignment	Exact m/z	Experimental m/z	% of abundance[a]		
			Quartz	Biotite	Orthoclase
[Purine + H]$^+$	121.0514	121.0471	37.3	30.2	32.5
[4-Azacytosine + Na]$^+$	135.0283	135.0230	40.2	29.0	30.8
[Purine + K]$^+$	159.0073	159.0026	39.4	24.5	36.1
[N(9)-formylpurine + K]$^+$	187.0022	187.0016	37.1	27.4	35.5
[2Cytosine + Na]$^+$	245.0763	245.0871	38.7	26.1	35.2

[a]Calculated as percentage of the total selected ion intensity summed for equal-area samples of each mineral type
Reproduced from [ref number, 2013] with permission from American Chemical Society

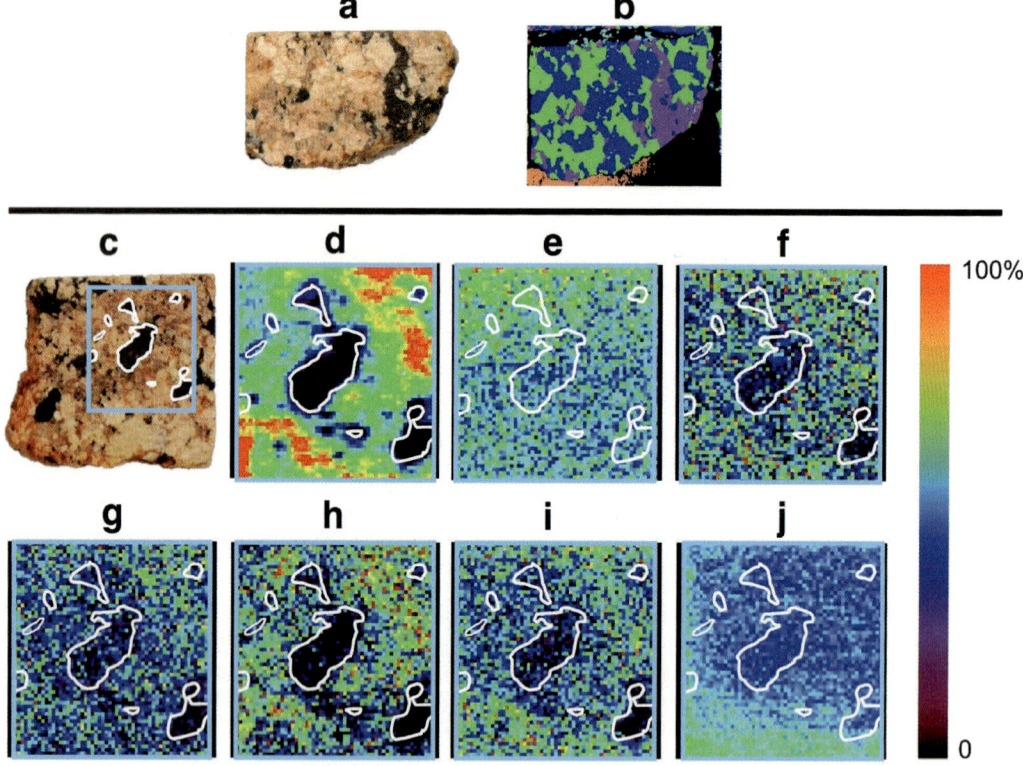

Fig. 2 (**a**) Optical image of a representative sample granite and (**b**) a corresponding μXRF image. (**c**) Optical image of reacted granite, (**d**) total ion image and selected ion images of (**e**) protonated purine (m/z 121), (**f**) sodiated 5-azacytosine (m/z 135), (**g**) potassiated purine (m/z 159), (**h**) potassiated N(9)-formylpurine (m/z 187), (**i**) sodiated cytosine dimer (m/z 245), and (**j**) phthalate ester (m/z 149) acquired with the area imaged outlined in *blue* and trends overlaid in *white*. Reproduced from [6] with permission from American Chemical Society

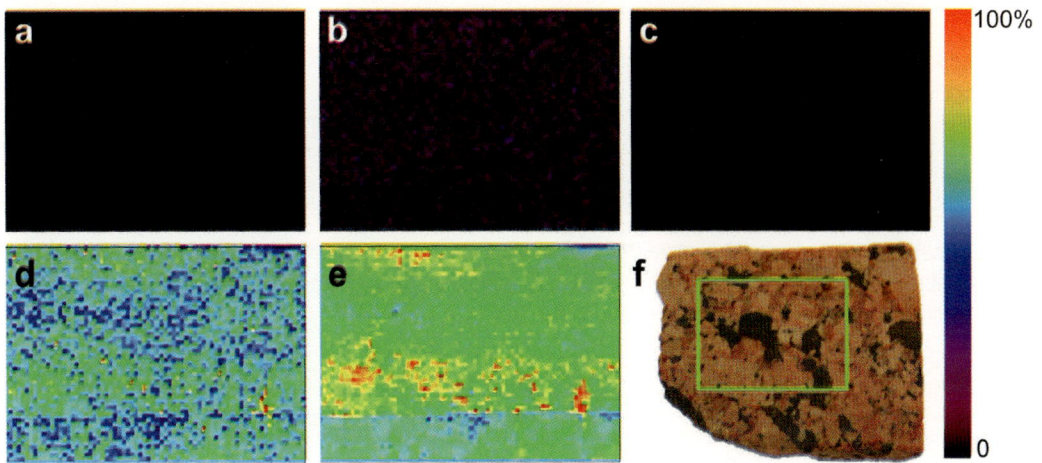

Fig. 3 Extracted ion images of (**a**) protonated purine, *m/z* 121; (**b**) protonated adenine, *m/z* 136; (**c**) potassiated *N*(9)-formylpurine, *m/z* 187; (**d**) solvent contaminant phthalate ester, *m/z* 149; and (**e**) total ion images obtained by DESI-MSI of a granite sample ((**f**), imaged area outlined in *green*) soaked in formamide. Reproduced from [6] with permission from American Chemical Society

To ensure that the observed differences in ion abundance across the surface are not artifacts of the DESI-MSI experiment, the following control experiments should be performed in addition to the control experiment for the model reaction.

3.8 Reaction Control (See Note 10)

1. With a new and clean granite sample, repeat Subheading 3.6, except do not place the sample in the oven.

2. Image the control granite sample according to Subheadings 3.1–3.4.

3. A set of chemical images acquired following this control reaction are shown in Fig. 3.

3.9 Transmission Control

1. When processing images of reacted granite surfaces, map the intensity of the solvent contaminant ion (Fig. 2j). Note any changes in the intensity throughout the images (*see* **Note 11**).

3.10 Adsorption Control

Given the large supernatant volume, it is necessary to ensure that the reaction product's spatial distribution across different mineral types does not result from the selective adsorption of chemicals in solution onto the surface.

1. Prepare a 25 µM solution of adenine in methanol (*see* **Note 12**).

2. Using a granite of similar composition, conduct the same cleaning and blank imaging outlined in Subheading 3.5.

3. Submerge granite sample in adenine solution for a minimum of 48 h at room temperature.

4. Remove the granite sample and proceed with imaging steps outlined in Subheading 3.3 (*see* **Note 13**). An example of this type of control image is shown in Fig. 4.

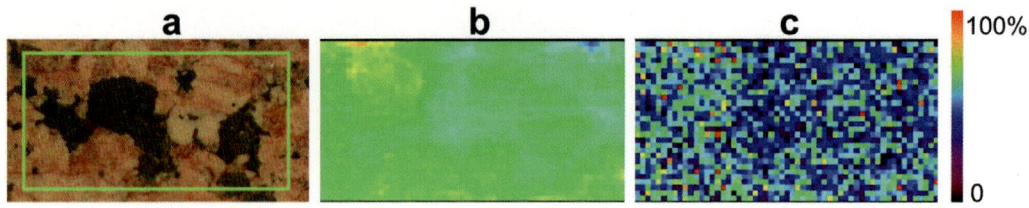

Fig. 4 Optical image (**a**) of granite sample that was soaked in adenine with area imaged by DESI-MSI outlined in *green*, total ion intensity image (**b**), and extracted ion image of protonated adenine, *m/z* 136 (**c**). Reproduced from [6] with permission from American Chemical Society

4   Notes

1. The specific granite in this study was composed of quartz (SiO_2 75 %), orthoclase ($KAlSi_2O_8$, 10 %), and biotite ($K(Mg, Fe^{2+})_3(Si_3Al)O_{10}(OH)_2$, 5 %). Because DESI uses a high-pressure spray, the granite should be able to withstand the pressure of the jet without dislodging particles into the mass spectrometer inlet; that is, powdery samples are not ideal.

2. The recommended DESI source position geometry is given in Subheading 3.1, **step 4**. However given slight variances in source configurations, this geometry should be optimized for each source and experiment. A detailed description of the geometry parameters that will affect DESI performance are detailed in ref. 13. This optimized source geometry provides suitable initial conditions for imaging, but must be further optimized when analyzing the granite surface as the substrate composition and analyte of interest will affect the maximum desorption and transfer of ions [16].

3. Due to the small size of the granite samples, we found that a level was not as accurate as the method just described.

4. For the data presented here, the total area of the granite surface area imaged was 10×10 mm (*x, y*) for this example granite. The stage moves in unidirectional scans with line stepping in order to sample the entire area of interest. The acquisition time for each individual row of the image is 50 s (0.87 min, which depends on the scan speed in the *x* dimension). With the desired 10 mm image height (*y* dimension), and based on the 200 μm step height, 50 rows must be acquired for a total image time of ~45 min. For different granites, images of a different size can easily be acquired depending on the sample. The total area of interest can easily be input into the Omni Spray control software along with the desired scan speed and line step. The software will then automatically calculate the acquisition time for each row.

5. The speed and step height within the motion parameters for the imaging stage should be selected based on the impact spot size as this ultimately determines imaging resolution. Therefore, if a different granite, solvent mixture, source geometry, etc. are used, these variables should be adjusted accordingly.

6. The high-pressure spray of the DESI provides an additional measure to clean the granite surface and the image acquired serves as a blank image.

7. The average literature-reported amount of reactants used in the formamide-mineral reactions is ~5 % mineral by weight with respect to the formamide [11]. Given the size of the granite samples used in the present studies, large amounts of formamide would be required to maintain the traditional reactant ratios. Therefore, the volume of formamide was reduced for practical reasons, but was still sufficient for satisfactory reaction yield. The substantial volume of formamide present during and after the reaction presents a challenge for DESI-MSI data interpretation due to adsorption effects. Initial formamide reaction conditions tested the feasibility of using smaller volumes of formamide to minimize this effect, however, this resulted in a very low abundance of the expected products on the surface such that imaging was not feasible. Therefore the larger (50 mL) volume in conjunction with the adsorption control experiment outlined in Subheading 3.10 was deemed appropriate.

8. The presence of methanol and water in the rinse solution used in preparation for imaging and the DESI spray may affect some reaction pathways, but is necessary to remove excess formamide to allow for the direct measurements of the surface-adsorbed species. This rinse step may also impact the spatial distribution of the reaction products on the surface. However, we found that the capability to image the products closest to the granite surface outweighed the adverse effects of the rinse step.

9. During the imaging process, the DESI spray will naturally gradually remove layers of organics on the surface. Subsequent repeated imaging of the same surface can provide depth profile information regarding the product distributions on the granite surface. The pressure of the spray will affect the spatial distribution of the products, however, we believe that the porosity of the granite will reduce this effect as opposed to other samples, and will be minimally significant depending on the mineral feature size within the granite.

10. This control experiment validates that the reaction products were formed as a result of the heating of the formamide in the presence of the granite sample and not contaminants that were present at the outset of the experiment or formed over the course of time.

11. During the imaging of the reacted granite samples, depending on solvent grade, impurities in the solvent may still be observed. These impurities can be identified through a reflective ESI [17–19] background spectrum acquired by spraying at a clean glass slide (versus the granite surface). For example, if using methanol, an ion at m/z 149 that originates from phthalate esters is often observed in ESI and DESI spectra. Theoretically, the concentration of this contaminant is constant in the DESI spray. Therefore changes in the intensity of this ion indicate a change in ion transmission due to differences in porosity and roughness of different mineral species on the surface.

12. Adenine was chosen because of its chemical similarity to the reaction products and would behave comparably in the presence of the granite.

13. Given that there is a homogenous distribution of the solvent contaminant from the transmission control (Subheading 3.5, **step 3**), any differences in the adenine abundance would indicate differences in chemical adsorption across the mineral.

References

1. Takáts Z, Wiseman JM, Gologan B, Cooks RG (2004) Mass spectrometry sampling under ambient conditions with desorption electrospray ionization. Science 306(5695):471–473. doi:10.1126/science.1104404

2. Venter A, Sojka PE, Cooks RG (2006) Droplet dynamics and ionization mechanisms in desorption electrospray ionization mass spectrometry. Anal Chem 78(24):8549–8555. doi:10.1021/ac0615807

3. Costa AB, Cooks RG (2007) Simulation of atmospheric transport and droplet-thin film collisions in desorption electrospray ionization. Chem Commun 38:3915–3917

4. Costa AB, Graham Cooks R (2008) Simulated splashes: elucidating the mechanism of desorption electrospray ionization mass spectrometry. Chem Phys Lett 464(1–3):1–8. doi:10.1016/j.cplett.2008.08.020

5. Nemes P, Vertes A (2007) Laser ablation electrospray ionization for atmospheric pressure, in vivo, and imaging mass spectrometry. Anal Chem 79(21):8098–8106. doi:10.1021/ac071181r

6. Bennett RV, Cleaves HJ, Davis JM, Sokolov DA, Orlando TM, Bada JL, Fernández FM (2013) Desorption electrospray ionization imaging mass spectrometry as a tool for investigating model prebiotic reactions on mineral surfaces. Anal Chem 85(3):1276–1279. doi:10.1021/ac303202n

7. Miyakawa S, James Cleaves H, Miller S (2002) The cold origin of life: A. Implications based on the hydrolytic stabilities of hydrogen cyanide and formamide. Orig Life Evol Biosph 32(3):195–208. doi:10.1023/a:1016514305984

8. Tian F, Kasting JF, Zahnle K (2011) Revisiting HCN formation in Earth's early atmosphere. Earth Planet Sci Lett 308(3–4):417–423. doi:10.1016/j.epsl.2011.06.011

9. Saladino R, Crestini C, Costanzo G, Negri R, Di Mauro E (2001) A possible prebiotic synthesis of purine, adenine, cytosine, and 4(3H)-pyrimidinone from formamide: implications for the origin of life. Bioorg Med Chem 9(5):1249–1253. doi:10.1016/s0968-0896(00)00340-0

10. Hudson JS, Eberle JF, Vachhani RH, Rogers LC, Wade JH, Krishnamurthy R, Springsteen G (2012) A unified mechanism for abiotic adenine and purine synthesis in formamide. Angew Chem Int Ed 51(21):5134–5137. doi:10.1002/anie.201108907

11. Saladino R, Claudia C, Giovanna C, Ernesto D (2004) Advances in the prebiotic synthesis of nucleic acids bases: implications for the origin of life. Curr Org Chem 8(15):1425–1443. doi:10.2174/1385272043369836

12. Costanzo G, Saladino R, Crestini C, Ciciriello F, Di Mauro E (2007) Formamide as the main building block in the origin of nucleic acids. BMC Evol Biol 7(Suppl 2):S1

13. Saladino R, Crestini C, Pino S, Costanzo G, Di Mauro E (2012) Formamide and the origin of life. Phys Life Rev 9(1):84–104. doi:10.1016/j.plrev.2011.12.002

14. Green FM, Stokes P, Hopley C, Seah MP, Gilmore IS, O'Connor G (2009) Developing repeatable measurements for reliable analysis of molecules at surfaces using desorption electrospray ionization. Anal Chem 81(6):2286–2293. doi:10.1021/ac802440w

15. Kertesz V, Van Berkel GJ (2008) Scanning and surface alignment considerations in chemical imaging with desorption electrospray mass spectrometry. Anal Chem 80(4):1027–1032. doi:10.1021/ac701947d

16. Crotti S, Traldi P (2009) Aspects of the role of surfaces in ionization processes. Comb Chem High Throughput Screen 12(2):125–136. doi:10.2174/138620709787315427

17. Van Berkel GJ, Ford MJ, Deibel MA (2005) Thin-layer chromatography and mass spectrometry coupled using desorption electrospray ionization. Anal Chem 77(5):1207–1215. doi:10.1021/ac048217p

18. Shin Y-S, Drolet B, Mayer R, Dolence K, Basile F (2007) Desorption electrospray ionization-mass spectrometry of proteins. Anal Chem 79(9):3514–3518. doi:10.1021/ac062451t

19. Douglass K, Jain S, Brandt W, Venter A (2012) Deconstructing desorption electrospray ionization: independent optimization of desorption and ionization by spray desorption collection. J Am Soc Mass Spectrom 23(11):1896–1902. doi:10.1007/s13361-012-0468-x

Chapter 9

Imaging of Plant Materials Using Indirect Desorption Electrospray Ionization Mass Spectrometry

Christian Janfelt

Abstract

Indirect desorption electrospray ionization mass spectrometry (DESI-MS) imaging is a method for imaging distributions of metabolites in plant materials, in particular leaves and petals. The challenge in direct imaging of such plant materials with DESI-MS is particularly the protective layer of cuticular wax present in leaves and petals. The cuticle protects the plant from drying out, but also makes it difficult for the DESI sprayer to reach the analytes of interest inside the plant material. A solution to this problem is to imprint the plant material onto a surface, thus releasing the analytes of interest from parts of their matrix while preserving the spatial information in the two dimensions. The imprint can then easily be imaged by DESI-MS. The method delivers simple and robust mass spectrometry imaging of plant material with very high success ratios.

Key words Mass spectrometry, MS imaging, DESI-MS, Plant metabolites, Sample preparation, Protocol

---

1 Introduction

While desorption electrospray ionization (DESI) mass spectrometry imaging is widely used for imaging of exogenous and endogenous compounds in sections of animal tissue [1], its use for imaging of plant material requires a set of new considerations. In particular, the layer of cuticular wax which protects leaves and petals from drying out presents a challenge not only in DESI imaging experiments, but also in imaging experiments using ionization techniques based on laser desorption [2]. This is probably also the reason why DESI imaging of plant material was not presented until 2011 (using the present indirect approach) [3] while DESI imaging of animal tissue had been performed since 2006 [4].

Direct, *non-imaging* DESI analysis of plant material is indeed possible and was early shown for a number of different plants [5, 6]. This is readily possible, despite the cuticle, because the DESI spray over time—typically over a few seconds—will work its way through

Lin He (ed.), *Mass Spectrometry Imaging of Small Molecules*, Methods in Molecular Biology, vol. 1203,
DOI 10.1007/978-1-4939-1357-2_9, © Springer Science+Business Media New York 2015

the cuticle and release the analytes for desorption and ionization. It is, however, much more critical in an imaging experiment that the analytes are readily accessible and that a stable signal is obtained. While in a non-imaging experiment one can always pick out the successful spectra, in an imaging experiment all spectra need to be successful, since unsuccessful spectra will cause irregular pixels in the image. Indeed, in some cases direct DESI imaging of plant materials is possible [7], but it requires very careful optimization and selection of operating parameters. Another way to perform direct DESI imaging of plant materials is of course to apply cryosectioning as part of the sample preparation, just as what is done for animal tissue, as shown for example in the imaging analysis of 50 μm thin sections of cassava tubers [8]. While this approach is useful in imaging of, e.g., seeds and thicker kinds of plant material, it is not feasible to perform cryosectioning of very thin objects such as leaves and petals. In those instances, it is advantageous to generate an imprint of the plant material and image the imprint subsequently. While paper has been used as imprinting material for DESI imaging of bacterial colonies [9], porous Teflon has proven to be an extremely suitable imprinting surface for indirect DESI imaging of plant material [3, 8, 10–13]. More recently, thin-layer chromatography (TLC) plates have been used in combination with solvent extraction and heating [14].

The advantage of indirect DESI imaging via a porous Teflon surface is the transfer of analytes from soft and irregular plant tissue, possibly coated with cuticular waxes to various extents, to the well-defined surface of porous Teflon. Porous Teflon has previously shown ideal properties for DESI analysis [15], combining the nonstick virtue of normal Teflon surfaces, which allows compounds to be easily desorbed, with the porosity ensuring that all sample is not washed away immediately by the DESI sprayer, as what is sometimes observed with smooth, nonporous Teflon surfaces. During the imprinting process, analytes dissolved in the plant juice will be transferred to the imprinting surface while maintaining their two-dimensional distribution, leaving the plant tissue and its possible matrix interferences behind. It has been shown that polar and nonpolar compounds are transferred equally well to the porous Teflon surface [3] (*see* **Note 1**). The obvious limitation of the approach is of course that the sample is "flattened" to a two-dimensional image and that all information about the third dimension is lost. This can in some cases be compensated for, e.g., by slicing of the sample prior to the imprinting process, such that each slice imprinted represents one layer in the original sample.

The procedures in indirect DESI imaging involve the production of an imprint and the subsequent imaging of the imprint, which is performed like any other DESI imaging experiment. The imprint is made by building a sandwich of the clean Teflon surface, the plant material, some tissue paper, and a rubber slice and pressing the entire sandwich in a vice for a few seconds.

2 Materials

1. Plant materials (*see* **Note 2**).

2. Porous Teflon (*see* **Note 3**).

3. Tissue paper (e.g., Kimcare medical wipes) for absorption of surplus plant juice.

4. Silicone rubber (reusable, 1 mm thick), to distribute the pressure over the entire piece of sample material.

5. Vice to pressure-produce the imprint.

6. Metal plates (e.g., 6 mm thick aluminum) to confine the sample.

7. Typical solvents used: Methanol or acetonitrile in combination with milli-Q water, possible with 1 % formic acid or ammonia added in order to improve the ionization in the positive or negative ion mode, respectively.

8. Thermo LTQ XL linear ion trap mass spectrometer (Thermo-Fisher Scientific, San Jose, CA, USA) or similar.

9. Custom-built DESI imaging ion source or the commercially available 2D DESI ion source from Prosolia (Indianapolis, IN, USA).

10. Camera or microscope equipped with a camera.

3 Methods

3.1 Imprinting of Plant Material (Fig. 1)

1. A piece of porous Teflon is cut in an appropriate size, e.g., 25×55 mm, and placed on a metal plate of similar size or larger.

2. The plant material is cut in the desired shape and photographed (for subsequent comparison between optical and MS images), either with a microscope equipped with a camera or with a standard camera. When a standard camera is used, it may be advantageous to place the plant material on a light plate.

3. The sample is placed on the porous Teflon (*see* **Note 4**), and covered by a few layers of tissue paper (Fig. 1a) (*see* **Note 5**).

4. A piece of silicone rubber is placed on top of the tissue paper (*see* **Note 6**) and another aluminum plate is placed on the top.

5. The whole sandwich is placed in a vice and a firm pressure is applied (Fig. 1b) (*see* **Note 7**). The imprint time can be as little as a few seconds or several minutes. In practice, the process is quite fast, and the pressure may be released upon a few seconds (*see* **Notes 8** and **9**).

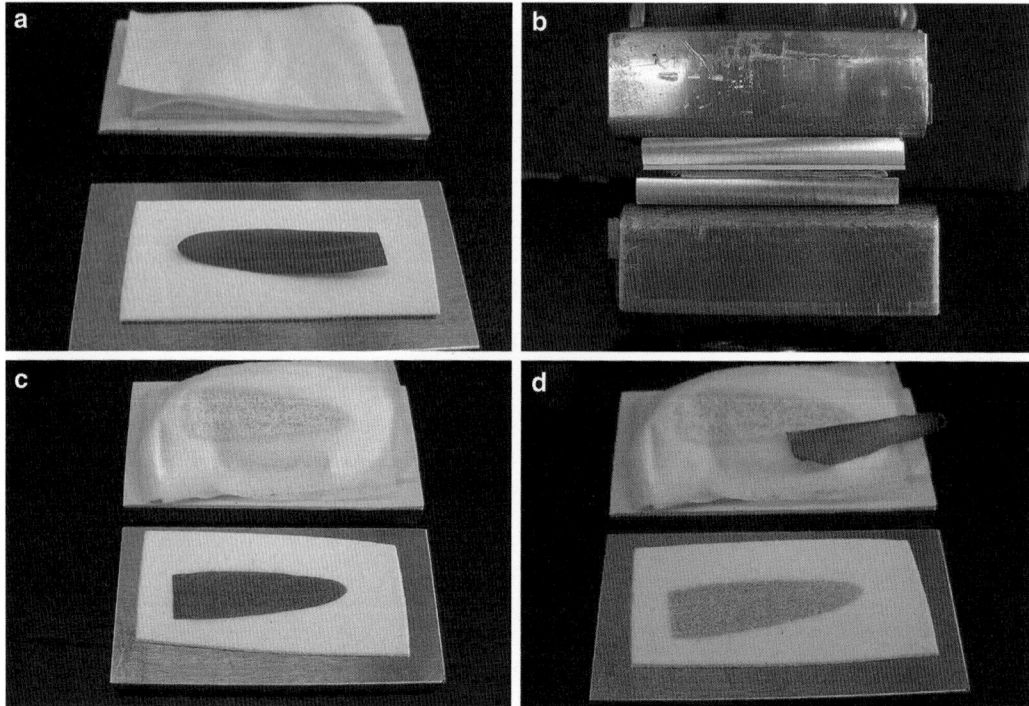

Fig. 1 The imprinting process, shown here for the imprinting of a barley leaf. (**a**) The leaf is placed on the clean, porous Teflon surface and covered with tissue paper and a rubber slice. (**b**) The sandwich is placed in a vice a pressed for a few seconds. (**c**) The sandwich is opened. (**d**) The imprint (here visible, but not necessarily always) is ready for imaging upon a few minutes of drying in a vacuum desiccator or under ambient conditions

6. The sandwich is released from the vice and opened. The plant tissue may stick to the porous Teflon (Fig. 1c) but can be removed with a pair of tweezers. The imprint (Fig. 1d) is dried under ambient conditions or in a vacuum desiccator (*see* **Notes 8 and 9**) for a few minutes and is subsequently ready for DESI analysis and imaging.

3.2 DESI Imaging of Imprinted Plant Material (See Note 8)

1. A photo is taken of the imprint, using a camera or a microscope, for subsequent comparison between the imprint and the MS images.

2. The imprint is placed in the DESI imaging ion source like any other sample. A mixture of, e.g., acetonitrile and water may be applied as spray solvent; the exact composition of the spray solvent is chosen and optimized according to the type of analyte to be imaged (*see* **Note 10**).

3. The spray geometry is optimized on an imprint not intended for imaging (*see* **Note 11**). The geometry optimization is relatively simple for the porous Teflon surface, and the settings are not nearly as critical as in, e.g., imaging of animal tissue section mounted on glass slides. Moreover, the success ratios with indirect DESI imaging are very high compared to other DESI imaging experiments.

4. The imaging experiment is carried out. If an imprint is considered to be unique to the extent that a similar one cannot be produced, yet images in both ion polarities are still desired, one may apply the displaced dual-mode imaging method [13] (*see* **Note 12**).

5. The DESI images are generated on a computer using the appropriate software.

4   Notes

1. Although the porous Teflon surface itself is extremely nonpolar, the method has been found not to discriminate significantly between polar and nonpolar compounds [3]. This was tested by comparing the DESI signal from an imprint of a capsule of the opium poppy (*Papaver somniferum*) with the electrospray ionization signal from a methanol extract of a similar capsule in a direct infusion ESI analysis. The mutual abundances between highly polar compounds such as morphine and less polar compounds such as papaverine were similar for both of the two techniques. This can most likely be ascribed to the plant juice, with all its contents, being squeezed into the pores of the Teflon surface by mechanical means. This results in a nonselective transfer of compounds, in contrast to an extraction process as known from, e.g., solid-phase extraction, which is much more chemically selective.

2. The plant material which is to be imaged must be fresh material with some amount of plant juice present which can transfer the analytes to the imprinting surface. The method is therefore not readily applicable for imaging of dried plant material.

3. Porous Teflon can be purchased, e.g., from Berghof (www.berghof.com, Eningen, Germany) which produces it on request with the desired specifications in terms of thickness, average pore size, and porosity (pore volume in %). For practical reasons, a thickness of 1.6 mm is preferable. The two other parameters can be combined as, e.g., 36 % porosity with 7 µm pore size or 47 % porosity with 16 µm pore size. Given the effective resolution of about 100 µm which is typically obtained in a DESI experiment, there does not seem to be much difference between images made with the two types of porous Teflon. In other DESI studies, porous Teflon with 25 µm pore size has been purchased from Small Parts Inc. (Miami Lakes, FL, USA) [15], or from Fluoro-Plastics (Philadelphia, PA, USA) [16].

4. In practice it is often desirable to imprint at least two pieces of the plant material next to each other, so that one piece can be used for test and optimization of the DESI signal, while the imprint intended for imaging is not touched until the actual start of the image acquisition.

5. The purpose of the tissue paper is to absorb the surplus plant juice which is not absorbed by the porous Teflon. In the absence of tissue paper, very wet and blurred imprints are typically obtained. The amount of tissue paper may be adjusted to the type of plant material which is to be imprinted.

6. The rubber slice may in some cases be omitted. Its purpose is to distribute the pressure when samples do not have a completely even thickness. In the absence of the rubber slice, one may for example find that the regions of a leaf lying close to the midrib of the leaf are not sufficiently imprinted, since the protrusion of the midrib may "protect" them from the pressure applied. In this case, the rubber slice—being more flexible than the metal plate—serves to apply the pressure evenly throughout the sample.

7. Ideally, one might wish for more a reproducible way to apply exactly the same pressure every time. However, in practice it works fine simply to tighten the vice firmly without further quantification of the forces involved. Indeed, in one paper with imprint imaging "a 10-mL round-bottom flask was used as a plunger to imprint the leaf sap onto the PTFE surface" [11], suggesting that the way the pressure was applied was not very critical.

8. It is preferable to perform the DESI imaging analysis immediately after the imprint is made, partly because of the risk of degradation of labile compounds over time and partly because of the possible diffusion which may occur over time (of course depending on the temperature and humidity under which the imprint is stored). In case the imaging of an imprint has to be postponed, one may store the imprint in a sealed box or plastic bag at −80 °C. On the day of analysis, the imprint is taken directly from the freezer to a vacuum desiccator for thawing out, in order to avoid condensation of ambient water on the cold imprint. In the case of leaves of *Hypericum perforatum*, it was found that the signals from hyperforin were significantly reduced in an imprint which has been stored for 6 days at room temperature compared to a similar imprint stored for 6 days at −80 °C.

9. In certain plant materials enzymatic reactions are initiated upon the cell disruption caused by the pressure exerted in the imprinting process. An example known from everyday life is the pressing of garlic, during which alliin is enzymatically converted by alliinase into allicin, responsible for the characteristic smell of fresh garlic. In the event that such enzymatic processes may occur, the imprinting must be performed very quickly (imprinting time 1–2 s) and the imprint immediately moved to a vacuum desiccator in order to rapidly dry the fresh imprint and thus quench the enzymatic reactions. That this is indeed possible was shown in the case of cyanogenic glucosides [8].

10. In general, for DESI analysis of porous Teflon surfaces quite low flow rate (1–2 µL/min) is used together with a modest nebulizer gas pressure (≤7 bar).

11. Although the pores in the surface prevent the sample from being used up immediately (as known for the nonporous Teflon surface), the DESI sprayer does have an impact on the imprint. The DESI analysis is thus not nondestructive and an imprint intended for an imaging experiment should therefore not be used for prior optimization of experimental conditions.

12. With this method, additional rows are inserted into the images and images in both polarities are recorded with the ion polarity switching between each row, yielding two separate images recorded with opposite polarities. Likewise, the method can be used for simultaneous acquisition of DESI images in full-scan and MS/MS mode.

References

1. Wu C, Dill AL, Eberlin LS et al (2013) Mass spectrometry imaging under ambient conditions. Mass Spectrom Rev 32:218–243

2. Cha SW, Zhang H, Ilarslan HI et al (2008) Direct profiling and imaging of plant metabolites in intact tissues by using colloidal graphite-assisted laser desorption ionization mass spectrometry. Plant J 55:348–360

3. Thunig J, Hansen SH, Janfelt C (2011) Analysis of secondary plant metabolites by indirect desorption electrospray ionization imaging mass spectrometry. Anal Chem 83:3256–3259

4. Wiseman JM, Ifa DR, Song QY et al (2006) Tissue imaging at atmospheric pressure using desorption electrospray ionization (DESI) mass spectrometry. Angew Chem Int Edit 45: 7188–7192

5. Talaty N, Takats Z, Cooks RG (2005) Rapid in situ detection of alkaloids in plant tissue under ambient conditions using desorption electrospray ionization. Analyst 130:1624–1633

6. Jackson AU, Tata A, Wu CP et al (2009) Direct analysis of Stevia leaves for diterpene glycosides by desorption electrospray ionization mass spectrometry. Analyst 134:867–874

7. Li B, Hansen SH, Janfelt C (2013) Direct imaging of plant metabolites in leaves and petals by desorption electrospray ionization mass spectrometry. Int J Mass Spectrom 348:15–22

8. Li B, Knudsen C, Hansen NK et al (2013) Visualizing metabolite distribution and enzymatic conversion in plant tissues by desorption electrospray ionization mass spectrometry imaging. Plant J 74:1059–1071

9. Watrous J, Hendricks N, Meehan M et al (2010) Capturing bacterial metabolic exchange using thin film desorption electrospray ionization-imaging mass spectrometry. Anal Chem 82:1598–1600

10. Li B, Bjarnholt N, Hansen SH et al (2011) Characterization of barley leaf tissue using direct and indirect desorption electrospray ionization imaging mass spectrometry. J Mass Spectrom 46:1241–1246

11. Müller T, Oradu S, Ifa DR et al (2011) Direct plant tissue analysis and imprint imaging by desorption electrospray ionization mass spectrometry. Anal Chem 83:5754–5761

12. Janfelt C, Nørgaard AW (2012) Ambient imaging mass spectrometry: a comparison of desorption ionization by sonic spray and electrospray. J Am Soc Mass Spectrom 23: 1670–1678

13. Janfelt C, Wellner N, Hansen HS et al (2013) Displaced dual-mode imaging with desorption electrospray ionization for simultaneous mass spectrometry imaging in both polarities and with several scan modes. J Mass Spectrom 48: 361–366

14. Cabral E, Mirabelli M, Perez C et al (2013) Blotting assisted by heating and solvent extraction for DESI-MS imaging. J Am Soc Mass Spectrom 24:956–965

15. Ifa DR, Manicke NE, Rusine AL et al (2008) Quantitative analysis of small molecules by desorption electrospray ionization mass spectrometry from polytetrafluoroethylene surfaces. Rapid Commun Mass Spectrom 22: 503–510

16. Nizzia JL, O'Leary AE, Ton AT et al (2013) Screening of cosmetic ingredients from authentic formulations and environmental samples with desorption electrospray ionization mass spectrometry. Anal Method 5:394–401

Imaging of Lipids and Metabolites Using Nanospray Desorption Electrospray Ionization Mass Spectrometry

Ingela Lanekoff and Julia Laskin

Abstract

Nanospray desorption electrospray ionization (nano-DESI) is an ambient ionization technique that uses localized liquid extraction for mass spectrometry imaging of molecules on surfaces. Nano-DESI enables imaging of ionizable molecules from a sample in its native state without any special sample pretreatment. In this chapter we describe the protocol for nano-DESI imaging of thin tissue sections.

Key words Nanospray desorption electrospray ionization (nano-DESI), Mass spectrometry imaging, Ambient ionization, Tissue sections, Lipids, Metabolites

1 Introduction

Nanospray desorption electrospray ionization (nano-DESI) is an ambient ionization technique [1] that enables sensitive mass spectrometry imaging of fully hydrated biological materials without any sample pretreatment. In nano-DESI experiments, a primary capillary, that continuously delivers a solvent to the sample, is brought in contact with a self-aspirating secondary capillary that removes the solvent from the sample, thereby forming a liquid bridge between the two capillaries called the nano-DESI probe. When the probe comes in contact with a sample surface, analyte molecules are desorbed from the surface into the solvent, transferred into the secondary capillary, and ionized by electrospray at a mass spectrometer inlet. Imaging is performed by continuously moving the sample under the probe in parallel lines while acquiring mass spectra [2]. A motorized XYZ stage controls the position of the sample holder and is programed to maintain the same distance between the sample and the probe to enable automated image acquisition [3]. The soft ionization in nano-DESI enables imaging of intact lipids and metabolites without interference of matrix

Lin He (ed.), *Mass Spectrometry Imaging of Small Molecules*, Methods in Molecular Biology, vol. 1203,
DOI 10.1007/978-1-4939-1357-2_10, © Springer Science+Business Media New York 2015

peaks. We have recently demonstrated highly sensitive imaging and quantification of nicotine in rat brain tissue sections at 0.35 fmoles/pixel by doping a deuterated nicotine standard into the nano-DESI solvent [4]. This approach also helped eliminate matrix effects, such as ion suppression, and enabled determination of a true nicotine image in brain tissue sections [4]. Although initially nano-DESI imaging has been developed on an Orbitrap XL (Thermo Fischer, San Jose) instrument, this imaging technique can be coupled to any mass spectrometer equipped with an electrospray ionization interface.

2 Materials

2.1 Solvents

A mixture of methanol:H_2O (9:1) has been successfully used to image lipids and metabolites in rat brain tissue sections, but a variety of solvent mixtures can be used as working solvents in nano-DESI imaging experiments (*see* **Note 1**). The solvent is delivered to the surface using a Hamilton syringe and a syringe pump (*see* **Note 2**).

2.2 Capillaries

1. Ceramic capillary cutter.

2. Polymer tubing cutter for 1/16″ PEEK tubing.

3. Fused silica capillaries come in a variety of sizes and several can be used for nano-DESI imaging. Here we describe the use of polyimide-coated fused silica capillaries (such as those from Polymicro Technologies L.L.C., Phoenix) with 50 μm inner diameter (ID) and 150 μm outer diameter (OD) to form the nano-DESI probe and a 250 μm × 360 μm (ID × OD) capillary as a solvent transfer line.

4. 1/16″ OD PEEK tubing with the ID matching the OD of the fused silica capillary.

5. Stainless steel and PEEK unions, fittings, and ferrules for connecting capillaries.

3 Methods

3.1 Fabrication of the Nano-DESI Probe Capillaries

3.1.1 Secondary Capillary

Fig. 1 shows a photograph of the fabricated capillaries.

1. Cut approximately 5 cm of a fused silica capillary (50 μm × 150 μm) and insert it into a ~0.5 cm long PEEK tubing (0.01″ × 0.0625″).

2. Place the capillary such that both sides extend evenly from the PEEK tubing and apply a small amount of loctite glue to attach the capillary to the PEEK tubing (*see* **Note 3**).

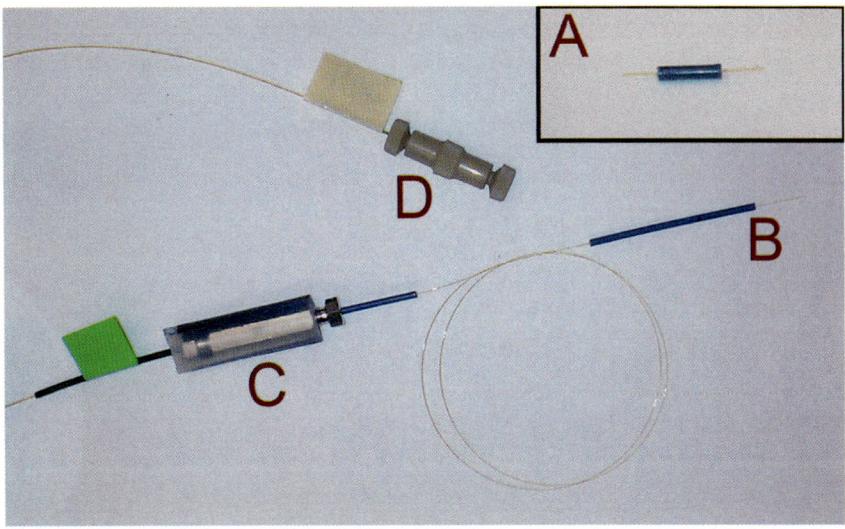

Fig. 1 Photograph of the primary and secondary capillaries after fabrication. (**A**) Secondary capillary, (**B**) tip of primary capillary, (**C**) stainless steel union, (**D**) PEEK union

3. Keeping the PEEK tubing in the center, cut the ends of the fused silica capillary to a final length of ~2 cm.

4. Examine the capillary under a microscope to make sure that it is cut evenly without any cracks or sharp points (*see* **Note 4**).

3.1.2 Primary Capillary

1. Cut 40–50 cm of fused silica tubing (50 μm × 150 μm) (small) and a piece of PEEK tubing (0.01 in. × 0.0625 in.) to ~2 cm.

2. Insert the fused silica capillary into the PEEK tubing and push it all the way through until approximately 2 cm of the capillary tip sticks out.

3. Glue the PEEK tubing and the capillary together (*see* **Note 3**) and make sure that the tip is nicely cut (*see* **Notes 4** and **5**).

4. Cut a second 40–50 cm piece of fused silica tubing (250 μm × 360 μm) (large).

5. Cut ~3 cm long pieces of 0.01 in. × 0.0625 in. and 0.02 in. × 0.0625 in. PEEK tubing and place these on the ends of the small and large capillary, respectively.

6. Add a ferrule and a male nut to both capillaries and join them using a stainless steel union (*see* **Notes 6** and **7**).

7. Add a micro tight sleeve (0.0155 in. × 0.025 in.) to the end of the large capillary, insert it into a PEEK fitting, and tighten the PEEK union.

8. Add a second PEEK fitting to the other side of the union and tighten it around a flat-headed Hamilton syringe filled with solvent. Propel the solvent through the primary capillary to make sure that it is properly tightened.

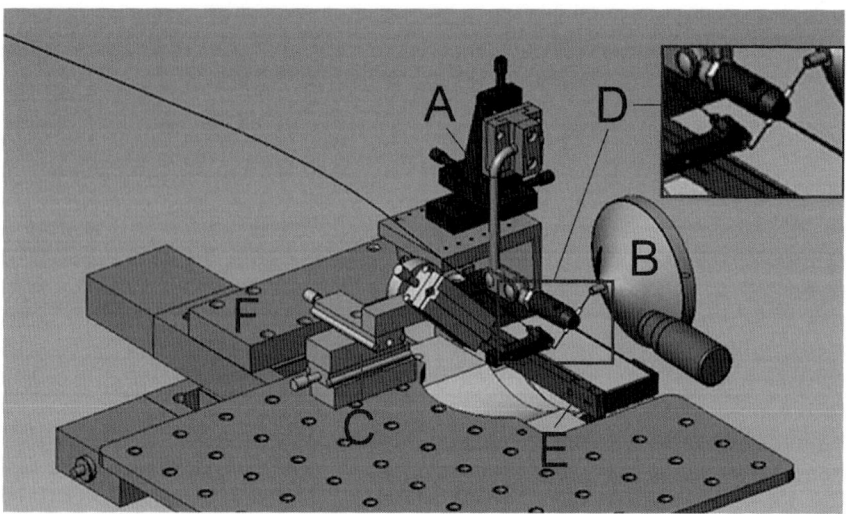

Fig. 2 Schematics of the nano-DESI setup. (**A**) Manual XYZ stage for positioning the secondary capillary, (**B**) mass spectrometer inlet, (**C**) manual XYZ stage for positioning the primary capillary, (**D**) nano-DESI probe, (**E**) sample holder attached to (**F**) computer-controlled XYZ stage

3.2 Fabricating a High-Resolution Nano-DESI Capillaries

1. A high-resolution nano-DESI probe is fabricated by pulling the fused silica capillaries. Pull the capillaries to the desired diameter, either using a laser puller or manually over a flame.

2. Follow Subheadings 3.1.1 and 3.1.2 to assemble the nano-DESI probe using pulled capillaries.

3.3 Setting Up the Nano-DESI Probe

1. Place a Hamilton syringe containing the working solvent into a syringe pump and attach the high-voltage supply to the stainless steel union using a crocodile clip.

2. Mount the secondary capillary onto a manual XYZ stage (Fig. 2a) and place it in front of the mass spectrometer inlet (Fig. 2b). Use the manual XYZ stage to move the secondary capillary close to the inlet so that the instrument vacuum can assist in drawing the liquid through the secondary capillary as shown in Fig. 3.

3. Mount the primary capillary onto a second manual XYZ stage (Fig. 2c) and position it such that the tip of the primary capillary touches the tip of the secondary capillary (Fig. 2d). This forms the nano-DESI probe (*see* **Notes 8** and **9**). The angle between the primary and secondary capillary should be around 90°. Fig. 4 shows a schematic drawing of the properly aligned nano-DESI probe.

4. Start the solvent flow using a syringe pump. Refine the relative position of the primary and the secondary capillary, as well as the position of the secondary capillary relative to the mass spectrometer inlet to ensure balanced solvent flow and stable electrospray.

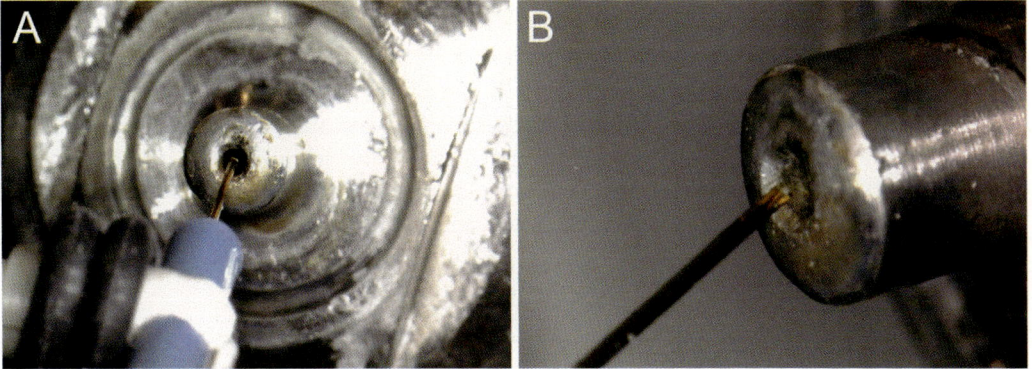

Fig. 3 The position of the secondary capillary in the inlet of the mass spectrometer. (**A**) Front view, (**B**) side view

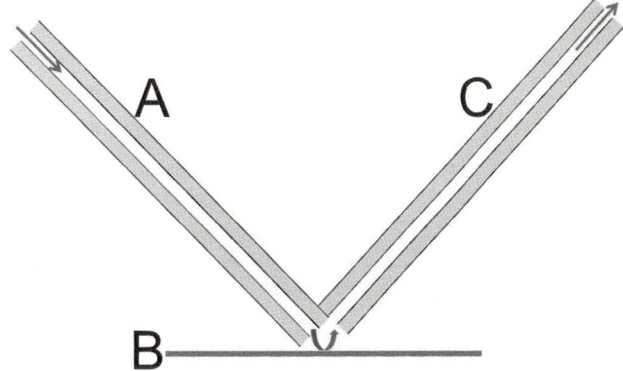

Fig. 4 Schematic drawing of a properly aligned nano-DESI probe. (**A**) Primary capillary, (**B**) sample surface, (**C**) secondary capillary

The solvent flow is balanced when the flow through the primary capillary matches the flow through the secondary capillary (*see* **Notes 9** and **10**).

3.4 Starting a Nano-DESI Imaging Experiment

1. Place a sample, attached to a regular glass microscope slide, into a sample holder (Fig. 2e) attached to a computer-controlled XYZ stage (Fig. 2f).

2. Move the sample holder toward the nano-DESI probe until the sample gets in contact with the probe. Observe the relative position of the sample and the probe using a digital camera. Ensure that both the solvent flow and the ion signal are stable (*see* **Note 11**).

3. Imaging is performed by scanning the sample under the nano-DESI probe in lines while acquiring mass spectra. Define the area to be analyzed, the speed at which the sample is scanned under the nano-DESI probe, and the line spacing (*see* **Note 12**).

4. Set up the distance between the nano-DESI probe and the sample and make sure to keep the distance constant throughout the experiment (*see* **Note 13**).

5. Program the acquisition of the mass spectrometer; on most systems this is performed by setting up a method and a sequence with contact closure. One line scan over the tissue corresponds to one sample in the sequence (*see* **Note 14**).

6. Start the acquisition.

4 Notes

1. The solvent mixture should be chosen based on several factors: (a) its efficiency at desorbing the molecules of interest, (b) its compatibility with electrospray ionization, and (c) its interaction with the sample surface. A polar solvent on a less polar surface is suitable since this will prevent the solvent from spilling onto the sample. Depending on the sample surface it might be appropriate to adjust the surface tension of the solvent.

2. It is advised to place the freshly mixed solvent into an ultrasonic bath for 5 min prior to filling the syringe and to remove air bubbles from the syringe.

3. When using glue to attach the PEEK tubing to the fused silica capillary, cut a small piece of an auxiliary capillary and dip it into a drop of glue. This will create a tiny droplet of glue on the auxiliary capillary. Slide the auxiliary capillary along the junction between the PEEK tubing and the capillary and let the glue dry. Make sure that the glue does not get close to the capillary tip.

4. If inexperienced with capillary cutting, a visual examination of the tip should be performed using a microscope.

5. A micro beveller system can be used to bevel the tip of the primary capillary delivering the solvent. This will allow positioning the nano-DESI probe closer to the surface without scratching the sample.

6. To avoid leakage when pressurizing the line, make sure to tighten the stainless steel union.

7. To limit the area of the stainless steel union with exposed high voltage, a silicon sleeve, made by cutting a tube of appropriate size open, should cover the metal junction.

8. Use cameras for visualization of the nano-DESI probe from the top and from the side view while positioning the capillaries. It is preferable to first find the XY position where the primary capillary touches the secondary capillary and then move the primary capillary down to adjust the z position.

9. Aligning the fused silica capillaries to form the nano-DESI probe is best achieved by placing the primary capillary slightly below the secondary capillary so that approximately half the secondary capillary is "resting" on the end of the primary capillary (Fig. 4). Once the capillaries are properly positioned, the liquid should flow from under the primary capillary to the sample surface and into the secondary capillary. The flow of solvent to the surface can be improved by positioning the primary capillary at a steeper angle.

10. To achieve a stable electrospray only small, if any, bubbles should flow through the secondary capillary. The placement of the secondary capillary relative to the inlet of the mass spectrometer should be adjusted until the signal is stable. If necessary this can be combined with adjustments of the flow rate; however it is advised to keep a constant flow rate throughout your experiment.

11. Even if the solvent flow and the ion signal are stable when the nano-DESI probe does not touch the sample, placing the probe on the slide sometimes affects the solvent flow into the secondary capillary and reduces the signal stability. These interferences often disappear after the probe is kept on the surface for a few minutes, but sometimes the nano-DESI probe is not set up optimally and adjustments of the relative capillary position must be made to obtain stable signal (*see* **Note 9**).

12. Make sure to have an appropriate space between the lines to avoid overlap between adjacent lines, leading to oversampling. The speed of the stage is typically kept between 10 and 40 µm/s.

13. The distance between the probe and the surface must be kept constant, preferably with the probe about 10–20 µm above the surface. For high spatial resolution experiments, this distance should be reduced to 5–10 µm. If the distance becomes too small the surface will be scratched by the nano-DESI probe and if the distance is too big the solvent will not touch the surface; hence no analytes will be desorbed or detected. Keeping a constant distance is best achieved by defining the plane of the surface before analysis and by programming the computer-controlled XYZ stage positioning the sample to move accordingly. Additionally, high-resolution digital cameras (we use long-working-distance Dino-lite cameras) can be used to constantly monitor the distance between the surface and the nano-DESI probe to manually adjust the z stage.

14. Set up a method using the time it will take for the nano-DESI probe to move in one line over the sample. Connect the contact closure to start acquisition when the stage starts moving.

Acknowledgments

The protocol described in this chapter was developed as part of the Chemical Imaging Initiative at Pacific Northwest National Laboratory (PNNL). The research was conducted under the Laboratory Directed Research and Development Program at PNNL, a multiprogram national laboratory operated by Battelle for the US Department of Energy (DOE) under Contract DE-AC05-76RL01830. The experiments were performed at EMSL, a national scientific user facility sponsored by the DOE's Office of Biological and Environmental Research and located at PNNL.

References

1. Roach PJ, Laskin J, Laskin A (2010) Nanospray desorption electrospray ionization: an ambient method for liquid-extraction surface sampling in mass spectrometry. Analyst 135: 2233–2236

2. Laskin J, Heath BS, Roach PJ et al (2012) Tissue imaging using nanospray desorption electrospray ionization mass spectrometry. Anal Chem 84:141–148

3. Lanekoff I, Heath BS, Liyu A et al (2012) Automated platform for high-resolution tissue imaging using nanospray desorption electrospray ionization mass spectrometry. Anal Chem 84:8351–8356

4. Lanekoff I, Thomas M, Carson JP et al (2013) Imaging nicotine in rat brain tissue by use of nanospray desorption electrospray ionization mass spectrometry. Anal Chem 85:882–889

Chapter 11

Electrospray Laser Desorption Ionization (ELDI) Mass Spectrometry for Molecular Imaging of Small Molecules on Tissues

Min-Zong Huang, Siou-Sian Jhang, and Jentaie Shiea

Abstract

The use of an ambient ionization mass spectrometry technique known as electrospray laser desorption ionization mass spectrometry (ELDI/MS) for molecular imaging is described in this section. The technique requires little or no sample pretreatment and the application of matrix on sample surfaces is unnecessary. In addition, the technique is highly suitable for the analysis of hard and thick tissues compared to other molecular imaging methods because it does not require production of thin tissue slices via microtomes, which greatly simplifies the overall sample preparation procedure and prevents the redistribution of analytes during matrix desorption. In this section, the ELDI/MS technique was applied to the profiling and imaging of chemical compounds on the surfaces of dry plant slices. Analyte distribution on plant slices was obtained by moving the sample relative to a pulsed laser and an ESI capillary for analyte desorption and post-ionization, respectively. Images of specific ions on sample surfaces with resolutions of 250 μm were typically created within 4.2 h for tissues with sizes of approximately 57 mm × 10 mm.

Key words ELDI, Ambient ionization, Molecular imaging, Plant slice

1 Introduction

Ambient mass spectrometry, an extension to MS, is a set of useful techniques for the analysis of samples under open-air conditions. The feature of ambient MS is its capacity for direct, rapid, real-time, and high-throughput analyses with little or no sample pretreatment. It allows for the analyses of a wide range of substances from various surfaces and matrices [1]. With the development of new variants, combinations, and hybrids, several different ambient ionization techniques have been developed and described [2–4]. Examples of these methods are desorption electrospray ionization (DESI) [5], direct analysis in real time (DART) [6], and electrospray laser desorption ionization (ELDI) [7]. These techniques use a variety of sampling, desorption, and ionization processes including bombardment of charged droplets and metastable atoms, thermal

Lin He (ed.), *Mass Spectrometry Imaging of Small Molecules*, Methods in Molecular Biology, vol. 1203, DOI 10.1007/978-1-4939-1357-2_11, © Springer Science+Business Media New York 2015

and laser desorption, and post-ionization in electrospray ionization (ESI) and atmospheric pressure chemical ionization (APCI) plumes, respectively. Several novel techniques have been developed over the last few years with many applications, in which one of the most important applications is imaging mass spectrometry with ambient ionization techniques because it is able to determine the spatial distributions of chemical constituents on sample surfaces. Desorption electrospray ionization (DESI) has been used to construct molecular images of several biological tissues, such as mouse pancreatic tissues, rat brain tissues, metastatic human liver adenocarcinoma tissues, human breast tissues, and canine abdominal tumor tissues [8–11]. The capability of electrospray laser desorption ionization (ELDI) for profiling and imaging several biological tissue slices and painting has been demonstrated [3, 12]. A similar technique known as laser ablation electrospray ionization (LAESI) has been used for chemical molecular imaging and depth profiling of water-rich leaf tissues for usage in metabolic studies [13]; the most recent use of this technique is for the simultaneous imaging of small metabolites and lipids in rat brain tissues in situ cell-by-cell imaging of plant tissues [14]. A technique using a low-temperature plasma probe (LTP) is also used for molecular imaging and has been used to analyze works of art including paintings and calligraphy [15]. Probe electrospray ionization (PESI), an ESI-based ambient ionization, has been shown to have potential for direct mouse brain imaging analysis in an atmospheric pressure environment [16]. In short, the variety of these techniques with respect to sampling, desorption, and ionization capabilities allows for the analysis of a broad range of samples using imaging mass spectrometry coupled with ambient ionization.

Here, the 2D molecular imaging of sample surfaces using electrospray laser desorption ionization mass spectrometry (ELDI/MS) is discussed in detail. The technique has been demonstrated to be useful in detecting proteins and small organic compounds on solids under ambient conditions [7, 17, 18]. Analyte molecules in the solid were desorbed using a pulsed laser and then post-ionized in an ESI plume. It is possible to obtain data on predominant chemical compounds on a particular area of the sample surface with the assistance of a stepper motor and laser desorption (LD) at a high spatial resolution. This procedure was extended to illustrate the application of ELDI/MS to the imaging of dry plant slices from *Oldham Elaeagnus*, an important traditional Chinese herb, with an emphasis on small-molecule detection. Line scans were obtained by continuously moving the sample between each predefined point. These line scans are combined into an array to produce a 2D image. Using the protocol provided here, it is possible to obtain point analyses for qualitative chemical analysis and line scans for quantitative analysis, after which images can be produced by combining the individual line scans.

2   Materials

1. Dry *Oldham Elaeagnus* was purchased from a local market.

2. Electrospray solution: Methanol and water (50 %, v/v) with 0.1 % acetic acid (*see* **Note 1**).

3. Fused silica capillary (100 μm i.d.).

4. Syringe pump (e.g., model 100 KD Scientific).

5. Double-sided tape.

6. A Bruker Esquire 3000 Plus ion trap mass spectrometer controlled by the *EsquireControl 5.2* data processing software was used; however, a mass spectrometer with a similar atmospheric pressure interface can be used.

7. 266 nm pulsed Nd:YAG laser (e.g., MINILITE I, Continuum Electro-Optics Inc., USA).

8. A laser power and energy meter (e.g., SOLO 2, Gentec-EO).

9. A three-axis precision automatic stage (e.g., DMVTEKS Co. Ltd, Taiwan) with a travel range of 10 cm.

10. Because the moving stage, laser system, and the mass spectrometer work independently, an *IMS_Control* software was created to control the three components; a contact signal received from the laser system and automatic XY moving stage controller triggers the start of data acquisition in *EsquireControl 5.2* at the beginning of each line scan. The data acquisition time per line established in *EsquireControl 5.2* must be equal to the time that is required to perform one line scan. Control software and integration of moving stage, laser system, and the mass spectrometer provided for customization are available upon request from Torbis Technology CO., LTD. (http://www.icpdas.com/distributors/country/torbis.htm).

11. Home-developed *ImagAnalysis v2.1* software, available upon request.

3   Methods

3.1   ELDI Imaging Experimental Setup (Fig. 1a)

1. An ESI emitter continuously sprayed solvent through a fused silica capillary at a flow rate of 150 μL/h using a syringe pump and a 2.5 mL syringe (*see* **Notes 2** and **3**). A nebulizing gas, commonly used in conventional ESI, was not used during ELDI processes. The ESI plume was directed toward the ion sampling orifice of the mass spectrometer (i.e., the ESI plume was parallel to the sample plate). The resulting analyte ions formed in the ESI plume were sampled into the mass analyzer through the ion sampling capillary. The electrospray needle,

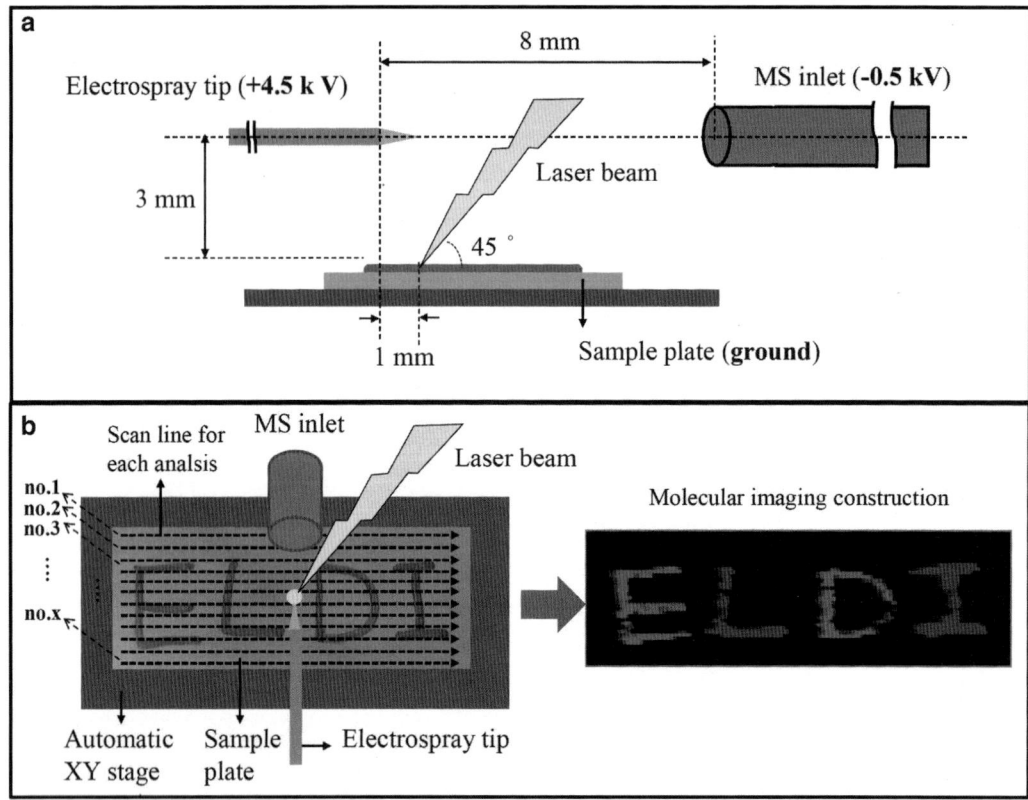

Fig. 1 (**a**) Graphic representation of the ELDI setup. The sample deposited on the stainless steel plate was positioned on the mobile sample stage and irradiated with a pulsed laser beam, where the laser beam was set behind the plane of the figure at an incident angle of 45°. The laser-ablated material was ionized in an electrospray solvent plume delivered through an electrospray capillary, where the resulting ions entered the mass spectrometer through the MS inlet tube. The distance between the ESI tip and the MS inlet tube was set as 8 mm, while the distance between the electrospray capillary and sample surface was set as 3 mm; the optimum location of the laser spot on the sample surface was positioned approximately 1 mm below the tip of the ESI capillary. (**b**) Schematic representation of the imaging experiment conducted using ELDI/MS. Each scan line on the sample resulted in a unique spectrum

the sample plate, and the sampling tube voltages were maintained at +4.5 kV, ground, and –0.5 kV, respectively (*see* **Note 4**).

2. A 266 nm pulsed Nd:YAG laser operating at a frequency of 10 Hz, a pulse energy of approximately 250 µJ (measured offline using a laser power and energy meter, SOLO 2, Gentec-EO), a pulse duration of 4 ns, and a spot size of approximately 250 µm was used for profiling and imaging analysis. The strongest ion signal was obtained at an incident laser angle of approximately 45° (*see* **Note 5**).

3. The geometry of the source was optimized to achieve an efficient mixing of ablated analytes with the ESI plume for maximum signal strength. The distance between the ESI tip and the MS

inlet tube was set as 8 mm, while the distance between the electrospray capillary and sample surface was set as 3 mm; the optimum location of the laser spot on the sample surface was positioned approximately 1 mm below the tip of the ESI capillary (*see* **Note 4**).

4. A three-axis precision automatic stage with a travel range of 10 cm was computer controlled when scanning the sample surface. While the laser beam irradiates the tissue slices, the sample stage is moved according to the laser beam at the speed of 200 μm/s in the longitudinal direction (X). The sample stage is further moved in the transverse direction (Y) upon computer-controlled positioning mechanism. Each scan line on the sample results in a unique spectrum (*see* **Note 6**).

5. The data acquisition programs rendered analysis times to the corresponding XY coordinates and converted the data sets into two-dimensional distributions. In-house software was used to produce contour plot images of the distribution of selected ions (*see* **Note 7**).

3.2 Preparation of Tissue Sections

1. The dry plant was cut into 2–5-mm-thick slices using a razor blade at room temperature. For *Oldham Elaeagnus*, the sample was sliced into thin sections of approximately $50 \times 60 \times 5$ mm (L×W×H) (*see* **Note 8**).

2. Place and fix the tissue sections onto sample plate using double-sided tape (*see* **Note 9**).

3. Dry tissue sections can be stored in the freezer for a few months

3.3 ELDI Imaging Experiments

1. The plant slice set on the sample plate was positioned on a homemade automated XY stage in front of the sampling capillary of a Bruker Esquire 3000 Plus ion trap mass spectrometer. For small-molecule analysis, the acquisition mass range was set from 50 to 250 *m/z*.

2. Use the *IMS_Control* software to set the moving speed of the stage and define the scanning area, and line scan number on the automatic XY stage software controller (*see* **Note 10**). For example, the sample stage is moved according to the laser beam at the speed of 200 μm/s along the *x*-axis within a defined area of 57×10 mm (L×W), each time at an increment of 250 μm in the transverse direction (Y) (*see* **Note 11**).

3.4 Data Acquisition

1. Create a sample list in the mass spectrometer acquisition software (*Bruker EsquireControl 5.2*). The total number of samples in the list is equal to the number of lines in the image. The last two characters of the file name should index the sequence of files (e.g., OE_01.yep, OE_02.yep, OE_03.yep, …, and OE_40.yep).

2. Make sure that the acquisition method contains the correct acquisition time for each line, e.g., 40 lines with an acquisition time of 285 s for each experiment.

3. Start the acquisition

3.5 Data Analysis

1. Before data analysis, convert the *Bruker EsquireControl 5.2* mass spectra files (.yep extensions) into format files (.ascii) using Bruker DataAnalysis software, in which the ascii files are required by home-developed *ImagAnalysis v2.1* software, which is available upon request.

2. The following instructions describe how to generate chemical images of *Oldham Elaeagnus* via *ImagAnalysis v2.1* software. (a) Open *ImagAnalysis v2.1*; (b) click on the LOAD DATA menu bar to load the converted file (.ascii extensions); (c) select the rainbow-colored scale and adjust the contrast of the image by selecting minimum and maximum values on the slide bars; (d) key in the m/z value (i.e., m/z 60.7, m/z 86.5, m/z 97.5, m/z 111.4, and m/z 137.5) displayed in the mass spectrum window; (e) click on the CREATE IMAGE menu bar to see an image of the distribution of small organic compounds from *Oldham Elaeagnus* surface; (f) copy the image and paste onto the organic photo of *Oldham Elaeagnus* using PowerPoint; and (g) repeat steps and the overlay chemical images from *Oldham Elaeagnus* surface should be observed (*see* **Note 7**).

4 Notes

1. The composition of the electrospray solution can influence the stability of the electrospray generated during ELDI analysis and the ability of the technique to detect particular analytes from the tissue matrix depending on the solubility of the analyte in the solvent system. The solvent composition is usually 50 % MeOH + 0.1 % acetic acid for most cases.

2. Ensure that the tip of the fused silica capillary is square and not burred or cracked. A burred or cracked tip will result in electrospray instability leading to irreproducible data.

3. Turn on the syringe pump and the high voltage for the electrospray. Make sure that the syringe contains enough solvent to acquire an image of the desired size. A 2.5 mL syringe filled with ESI solution is suggested and will typically last for 6 h with a flow rate of 150 µL/h. Let the electrospray stabilize for 10 min before starting analysis.

4. The voltage will depend on the composition of ESI solution, geometry of MS inlet, and the distances between the ESI tip, the MS inlet tube, and the sample plate. The high voltage is typically set from +4 kV to +6 kV for positive ion scan mode.

5. The laser energy will depend on the sample materials. The laser energy is typically set from 250 μJ to 1 mJ (*Warning*). Avoid eye or skin exposure to direct or scattered radiation. Safety goggles should be worn when performing ELDI experiments.

6. As shown in Fig. 2, molecular imaging analysis of a dry plant slice was completed in a few hours under a full scan in positive ion mode of mass spectrometry. The extracted ion chromatograms corresponding to m/z 86.5, m/z 111.4, m/z 97.5, and m/z 60.7 were acquired from a single scan line of different surfaces of an *Oldham Elaeagnus* slice.

7. As shown in Fig. 3, a typical set of molecular images on dry *Oldham Elaeagnus* slice were obtained by using ELDI/MS. The images were recorded at m/z 97.5, m/z 137.5, m/z 60.7, m/z 111.4, and m/z 86.5. The experimental procedures are extremely simple when compared to other MS imaging techniques because little or no sample pretreatment is required and matrix application of matrix on the sample surface is unnecessary. In addition, the technique is highly suitable for the analysis of hard and thick tissues without microtome production of tissue slices. In general, this protocol can be applied when an ELDI ion source and an automatic XY stage are coupled with a mass analyzer. The analytical steps described here are general and can be used in other applications involving ELDI (*see* **Note 12**).

8. Perform molecular imaging on a relatively hard and thick plant surface where a micrometer-scale sample slice cannot be obtained for the high degree of texture and fragile of dry plant tissue. The best thickness for this kind of tissue would be 2–5 mm. Please note that the surface of tissue should be smooth and crackles.

9. Adhere plant tissue samples to metal, plastic, or glass plates using double-sided tape.

10. To obtain the correct setup for the required image resolution, divide the sample area of the tissue by the different scan numbers according to sample size (*see* Fig. 1b). For example, *Oldham Elaeagnus* with a defined scanning width of 10 mm will result in 40 scan number (each time at an increment of 250 μm in the transverse direction).

11. Calculate the acquisition time based on the distance of each line scan and the moving speed of the automatic XY stage. For example, *Oldham Elaeagnus* with a defined scanning length of 57 mm and a moving speed of 200 μm/s along the *x*-axis will result in a duration of 285 s for each line scan. About 1,425 data points (mass spectra) will be obtained for each line scan while a Bruker Esquire 3000 plus ion trap mass spectrometer with a scan rate of 200 ms.

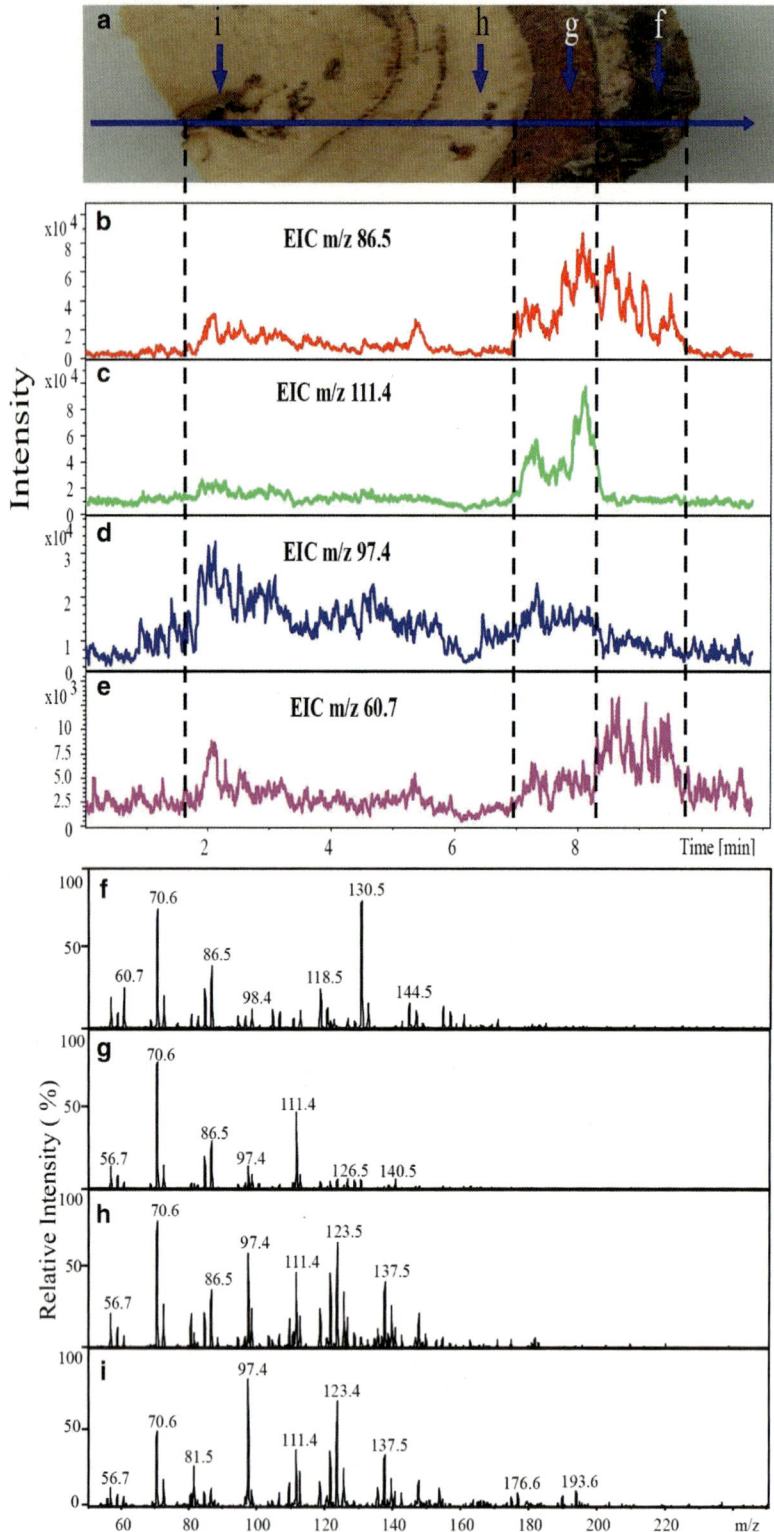

Fig. 2 (**a**) Photograph of an *Oldham Elaeagnus* slice, where locations of scanning lanes are indicated by the *dotted lines*; (**b–e**) extracted ion chromatograms from a full scan in positive ion mode corresponding to *m/z* 86.5, *m/z* 111.4, *m/z* 97.5, and *m/z* 60.7, respectively; (**f–i**) ELDI mass spectra from the surface of an *Oldham Elaeagnus* slice. Analytes were detected on the surface of a slice of *Oldham Elaeagnus* tree bark using ELDI/MS under ambient conditions without sample pretreatment. The location of different sampling spots was indicated by the *arrow* shown in panel (**a**)

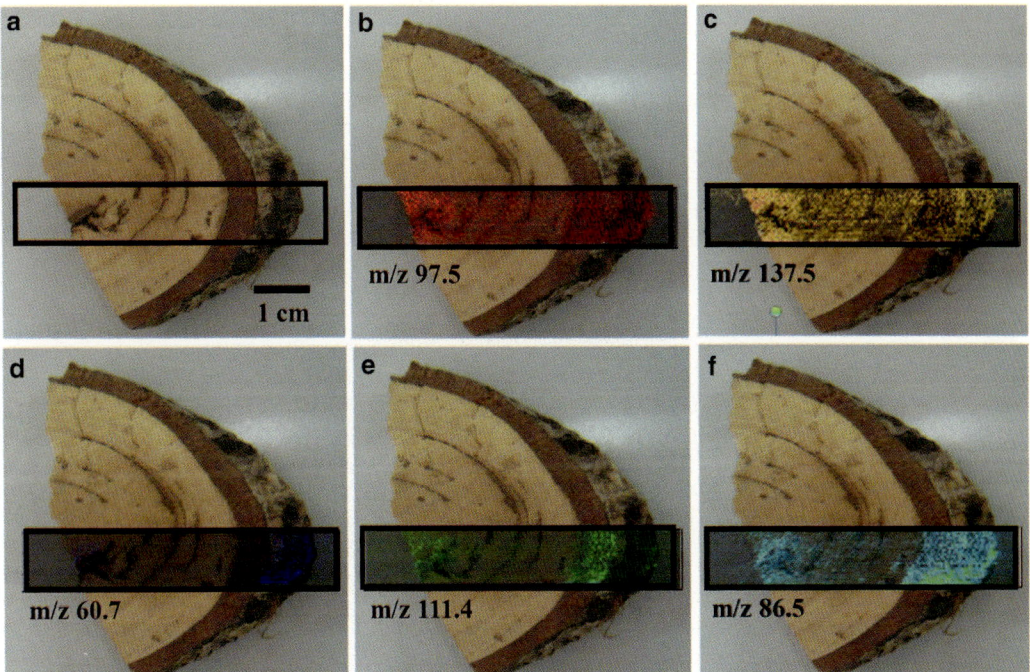

Fig. 3 (**a**) Photograph of a slice of *Oldham Elaeagnus* bark and the scanning area with a size of 57 × 10 × 4 mm (L × W × H); (**b–f**) molecular images corresponding to *m/z* 97.5, *m/z* 137.5, *m/z* 60.7, *m/z* 111.4, and *m/z* 86.5 from the *Oldham Elaeagnus* sample were recorded using ELDI/MS under ambient conditions

12. All operations should be conducted under appropriate safety protocols especially with respect to exposure at the operator and decontamination of laboratory equipment used in these studies.

References

1. Cooks RG, Ouyang Z, Takats Z, Wiseman JM (2006) Ambient mass spectrometry. Science 311:1566–1570

2. Huang MZ, Yuan CH, Cheng SC, Cho YT, Shiea J (2010) Ambient ionization mass spectrometry. Annu Rev Anal Chem 3:43–65

3. Fernandez FM, Harris GA, Galhena AS (2011) Ambient sampling/ionization mass spectrometry: applications and current trends. Anal Chem 83:4508–4538

4. Van Berkel GJ, Pasilis SP, Ovchinnikova O (2008) Established and emerging atmospheric pressure surface sampling/ionization techniques for mass spectrometry. J Mass Spectrom 43:1161–1180

5. Takats Z, Wiseman JM, Gologan B, Cooks RG (2004) Mass spectrometry sampling under ambient conditions with desorption electrospray ionization. Science 306:471–473

6. Cody RB, Laramee JA, Durst HD (2005) Versatile new ion source for the analysis of materials in open air under ambient conditions. Anal Chem 77:2297–2302

7. Shiea J, Huang MZ, Hsu HJ, Lee CY, Yuan CH, Beech I, Sunner J (2005) Electrospray-assisted laser desorption/ionization mass spectrometry for direct ambient analysis of solids. Rapid Commun Mass Spectrom 19:3701–3704

8. Cooks RG, Eberlin LS, Ifa DR, Wu C (2010) Three-dimensional visualization of mouse brain by lipid analysis using ambient ionization mass spectrometry. Angew Chem Int Ed Engl 49:873–876

9. Cooks RG, Wiseman JM, Ifa DR, Song QY (2006) Tissue imaging at atmospheric pressure using desorption electrospray ionization (DESI) mass spectrometry. Angew Chem Int Ed Engl 45:7188–7192

10. Gerbig S, Golf O, Balog J, Denes J, Baranyai Z, Zarand A, Raso E, Timar J, Takats Z (2012) Analysis of colorectal adenocarcinoma tissue by desorption electrospray ionization mass spectrometric imaging. Anal Bioanal Chem 403: 2315–2325

11. Dill AL, Ifa DR, Manicke NE, Zheng OY, Cooks RG (2009) Mass spectrometric imaging of lipids using desorption electrospray ionization. J Chromatogr B 877:2883–2889

12. Huang MZ, Cheng SC, Jhang SS, Chou CC, Cheng CN, Shiea J, Popov IA, Nikolaev EN (2012) Ambient molecular imaging of dry fungus surface by electrospray laser desorption ionization mass spectrometry. Int J Mass Spectrom 325:172–182

13. Nemes P, Barton AA, Li Y, Vertes A (2008) Ambient molecular imaging and depth profiling of live tissue by infrared laser ablation electrospray ionization mass spectrometry. Anal Chem 80:4575–4582

14. Shrestha B, Patt JM, Vertes A (2011) In situ cell-by-cell imaging and analysis of small cell populations by mass spectrometry. Anal Chem 83:2947–2955

15. Zhang SC, Liu YY, Ma XX, Lin ZQ, He MJ, Han GJ, Yang CD, Xing Z, Zhang XR (2010) Imaging mass spectrometry with a low-temperature plasma probe for the analysis of works of art. Angew Chem Int Ed Engl 49:4435–4437

16. Hiraoka K, Chen LC, Yoshimura K, Yu Z, Iwata R, Ito H, Suzuki H, Mori K, Ariyada O, Takeda S, Kubota T (2009) Ambient imaging mass spectrometry by electrospray ionization using solid needle as sampling probe. J Mass Spectrom 44:1469–1477

17. Huang MZ, Hsu HJ, Lee LY, Jeng JY, Shiea LT (2006) Direct protein detection from biological media through electrospray-assisted laser desorption ionization/mass spectrometry. J Proteome Res 5:1107–1116

18. Huang MZ, Hsu HJ, Wu CI, Lin SY, Ma YL, Cheng TL, Shiea J (2007) Characterization of the chemical components on the surface of different solids with electrospray-assisted laser desorption ionization mass spectrometry. Rapid Commun Mass Spectrom 21:1767–1775

Chapter 12

Automated Cell-by-Cell Tissue Imaging and Single-Cell Analysis for Targeted Morphologies by Laser Ablation Electrospray Ionization Mass Spectrometry

Hang Li, Brian K. Smith, Bindesh Shrestha, László Márk, and Akos Vertes

Abstract

Mass spectrometry imaging (MSI) is an emerging technology for the mapping of molecular distributions in tissues. In most of the existing studies, imaging is performed by sampling on a predefined rectangular grid that does not reflect the natural cellular pattern of the tissue. Delivering laser pulses by a sharpened optical fiber in laser ablation electrospray ionization (LAESI) mass spectrometry (MS) has enabled the direct analysis of single cells and subcellular compartments. Cell-by-cell imaging had been demonstrated using LAESI-MS, where individual cells were manually selected to serve as natural pixels for tissue imaging. Here we describe a protocol for a novel cell-by-cell LAESI imaging approach that automates cell recognition and addressing for systematic ablation of individual cells. Cell types with particular morphologies can also be selected for analysis. First, the cells are recognized as objects in a microscope image. The coordinates of their centroids are used by a stage-control program to sequentially position the cells under the optical fiber tip for laser ablation. This approach increases the image acquisition efficiency and stability, and enables the investigation of extended or selected tissue areas. In the LAESI process, the ablation events result in mass spectra that represent the metabolite levels in the ablated cells. Peak intensities of selected ions are used to represent the metabolite distributions in the tissue with single-cell resolution.

Key words Mass spectrometry, Imaging, Single-cell analysis, Cell-by-cell imaging, Metabolites, Tissue imaging, Molecular imaging

1 Introduction

Mass spectrometry imaging (MSI) is a rapidly emerging technique that enables the visualization of two- and three-dimensional distributions of metabolites, lipids, and proteins in biological tissues [1–5]. It complements the capabilities of conventional molecular histology by directly correlating molecular distributions with the histological features obtained from microscopy [1]. Established methods, such as MSI by matrix-assisted laser desorption ionization (MALDI) [6], and novel methods based on atmospheric pressure ionization, such as desorption electrospray ionization (DESI) [7]

Lin He (ed.), *Mass Spectrometry Imaging of Small Molecules*, Methods in Molecular Biology, vol. 1203,
DOI 10.1007/978-1-4939-1357-2_12, © Springer Science+Business Media New York 2015

and laser ablation electrospray ionization (LAESI) [8, 9], have demonstrated their ability to image diverse biological tissues. In most existing studies, MSI is performed by sampling based on a pre-defined gridding algorithm that follows a geometric pattern and ignores the cellular structure of the tissue [6, 7, 9–11]. As a consequence, molecular information from multiple cells may be captured together and cellular differences can be obscured [12].

As cells are the structural and functional units within a tissue, they represent the natural selection for pixels and voxels for two- and three-dimensional molecular imaging, respectively. To utilize this concept in MSI, analysis methods are needed for single-tissue-embedded cells. Optical fiber-based laser ablation in LAESI-MS can sample individual cells in their native environment [13, 14]. Using this method, cell-by-cell molecular imaging of metabolites in plant epidermal tissue was demonstrated [15]. Imaging was performed by manually moving the sample stage between cells. Manual control of the stage movement, however, is not feasible for the analysis of numerous single cells or selected cell types in extended tissue areas. Image processing combined with automating cell-by-cell imaging can overcome these limitations and provide a stable and efficient way to locate cells for analysis, thus enabling the investigation of extended or selected tissue areas. In this chapter, we present a protocol for automated cell-by-cell imaging of biological tissues using optical fiber-based laser ablation in LAESI-MS.

The procedure consists of the following major steps. Initially the tissue is inspected by an optical microscope and an image of the relevant area is captured. This image is processed to identify the coordinates of the cell centroids. These coordinates are used to program an automated translation stage that presents the individual cells one by one to the etched end of the optical fiber for ablation by the laser pulses. The plumes from the cell ablations are ionized by an electrospray and the produced ions are detected by a mass spectrometer. The peak intensities in the recorded mass spectra are used to create metabolite-specific ion intensity maps using the cells as pixels.

2   Materials

2.1   Reagents and Chemicals

1. Electrospray solution for ionization in positive ion mode: methanol with 0.1 % acetic acid (v/v):HPLC-grade water (1:1, v/v).

2. 1-Methyl-2-pyrrolidinone.

3. Germanium oxide (GeO_2)-based glass optical fiber (450 μm core diameter, Infrared Fiber Systems Inc., Silver Spring, MD, USA) (*see* Subheading 3.1).

4. Sapphire scribe.

5. Bare fiber chuck (BF300, Siskiyou Corporation, Grants Pass, OR, USA).

6. A translation stage (Thorlabs, Newton, NJ, USA).

7. Nitric acid: 1.0–2.4 % (*v/v*) reagent grade.

2.2 Biological Samples

1. Plant tissues such as Easter lily (*Lilium longiflorum*), leek (*Allium ampeloprasum*) and onion (*Allium cepa*) bulb.

2. Microtome knife blades for tissue excising.

3. Pre-cleaned microscope glass slides.

2.3 Cell Coordinate Recognition

1. An upright optical microscope (BX51, Olympus America Inc., Center Valley, PA, USA) was used for the imaging of cells in plant tissues.

2. Software used for image processing and cell coordinate measurements included ImageJ (Version 1.40 g, National Institute of Health, Bethesda, MD, USA) and MetaMorph for Olympus (Version 7.5.6.0, Olympus America Inc., Center Valley, PA, USA).

2.4 Single-Cell LAESI-MS

2.4.1 Microscope Visualization System

1. A homebuilt long-distance microscope, comprising a 7× precision zoom optic (Edmund Optics, Barrington, NJ, USA), a 2× infinity corrected objective lens (M Plan Apo 2×, Mitutoyo, Kanagawa, Japan), and a digital camera (Marlin F131, Allied Vision Technologies, Stadtroda, Germany), was mounted to provide a top view of the sample and the fiber tip and visualize the targeting and laser ablation of individual cells.

2. A similar long-distance microscope, built with the 7× precision zoom optic (Edmund Optics, Barrington, NJ, USA), a 5× infinity corrected objective lens (M Plan Apo 5×, Mitutoyo, Kanagawa, Japan), and a digital camera (Marlin F131, Allied Vision Technologies, Stadtroda, Germany), was positioned at a shallow angle to the sample (side view) to monitor the distance between the fiber tip and the sample surface. Positioning the fiber tip approximately a tip diameter away from the surface enabled effective laser ablation without fiber breakage.

2.4.2 Electrospray

1. The electrospray assembly components included an emitter (i.d. 50 µm, MT320-50-5-5, New Objective, Woburn, MA, USA), a metal union with a conductive perfluoroelastomer ferrule, fittings, a tubing sleeve, a needle port, and a fused silica capillary (IDEX Health and Science, Oak Harbor, WA, USA). High voltage was supplied by a regulated power supply (PS350, Stanford Research Systems, Sunnyvale, CA, USA) (*see* **Note 1**).

2. A syringe pump (Physio 22, Harvard Apparatus, Holliston, MA, USA) and a 500 µL syringe (Hamilton, Reno, NV, USA) were utilized to pump the electrospray solvent through the emitter.

<table>
<tr><td>

2.4.3 Laser Pulse Delivery

</td><td>

1. A mid-IR optical parametric oscillator, driven by a Q-switched Nd:YAG laser (Vibrant IR, Opotek, Carlsbad, CA, USA), produced 5 ns pulses at 2,940 nm wavelength with a repetition rate of 10 Hz (*see* **Note 2**).

2. The blunt end of fiber was mounted on a miniature 5-axis translator (BFT-5, Siskiyou, Grants Pass, OR, USA) with a bare fiber chuck. The sharpened end was manipulated by a micromanipulator (MN-151, Narishige, Tokyo, Japan).

3. A plano-convex CaF_2 lens with a focal length of 76.2 mm (Infrared Optical Products, Farmingdale, NY, USA) was used to focus the laser beam onto the blunt end of the optical fiber. Care was taken that the laser pulses filled out the entire cross section of the core and did not damage the blunt fiber end.

</td></tr>
</table>

2.5 Molecular Imaging and Data Analysis

1. An orthogonal acceleration time-of-flight mass spectrometer (Q-TOF Premier, Waters, Milford, MA, USA) was used to acquire mass spectra. The commercial electrospray source was replaced by our fiber-based LAESI source.

2. A three-axis translation stage with motorized actuators and a stage controller (LTA-HS, Newport, Irvine, CA, USA) provided the basis of accurate sample stage movement.

3. The stage movement scanning program, to target the centroids of individual cells based on their coordinates, was written in house using a visual programming platform (LabView, National Instruments, Austin, TX, USA).

4. Software for data analysis and molecular image processing included a scientific visualization package (Origin 8.0, Origin Lab, Northampton, MA, USA), and image processing programs (Photoshop 7.0, Adobe Systems Inc., San Jose, CA, USA, and ImageReady 7.0, Adobe Systems Inc., San Jose, CA, USA).

3 Methods

Automated cell-by-cell imaging by LAESI-MS relies on cell recognition and cell addressing derived from optical microscope images of the sample. For the ablation of individual cells, the centroid coordinates of targeted cells were determined through image processing. A stage-control program was developed to take these coordinates as input and sequentially position individual cells under the fiber tip for ablation. The ablation events for each cell resulted in a mass spectrum that reflected numerous molecular components of the cell. Peak intensities for selected ions were determined to build a false color image of cell-by-cell metabolite distributions in the tissue. Alternatively, cells of different morphologies were selectively targeted for analysis.

3.1 Preparation of Optical Fiber with a Sharpened End

1. 1-Methyl-2-pyrrolidinone was heated to 130–150 °C in a small beaker. A coated germanium oxide (GeO_2)-based glass optical fiber was dipped into the heated solvent for a minute until its plastic coating turned soft and started to peel off. The fiber was removed from the solvent, and quickly dipped into methanol to wash off the coating. Lint-free tissue was used to wipe off any remaining coating.

2. Both ends of the fiber were cleaved using a sapphire scribe (KITCO Fiber Optics, Virginia Beach, VA, USA) by scoring and gently snapping the fiber.

3. For etching, one end of the fiber was held using a bare fiber chuck and positioned vertically using a translation stage. The mounted fiber end was dipped into 1.0–2.4 % (v/v) reagent-grade nitric acid to a vertical depth of 0.3–0.5 mm. After ~15 min, when the etching was completed, the fiber tip automatically detached from the acid surface. The etched sharp fiber tip was rinsed with deionized water.

3.2 Cell Coordinate Recognition

1. A microtome knife blade and fine tweezers were used to cut and peel off the abaxial or adaxial epidermal layer of plant tissue. The peeled layer was mounted on a clean microscope slide.

2. A reference marker (i.e., a dot by a waterproof marker pen) was placed on the backside of microscope slide, and the plant (e.g., *L. longiflorum*) epidermis was observed under the upright microscope. A unique cell or feature on the tissue was selected with reference to the marker, and a recognizable point in it was defined as the origin of the coordinate system for the selected tissue area.

3. The defined origin was positioned at the top left corner of the field of view in the microscope. Images of the selected tissue area were captured at different magnifications and with different imaging modes (*see* **Note 3**).

4. Image processing software, e.g., ImageJ or MetaMorph for Olympus, was used to threshold and binarize the microscope image, and accentuate the cell edges. Objects, corresponding to cells, were identified in these binarized images and integrated morphometry analysis was performed to determine the centroids of each object (*see* **Note 4**). The centroid coordinate dataset was exported into the stage-control program for addressing the cells. The cell coordinate recognition process for the cell-by-cell imaging of the adaxial epidermis of *L. longiflorum* leaf is shown in Fig. 1a–c. Figure 2a–c shows the differentiation of two cell types (pavement cells and guard cells) in the abaxial epidermis of an *L. longiflorum* leaf and captures the coordinates corresponding to one of them (guard cells).

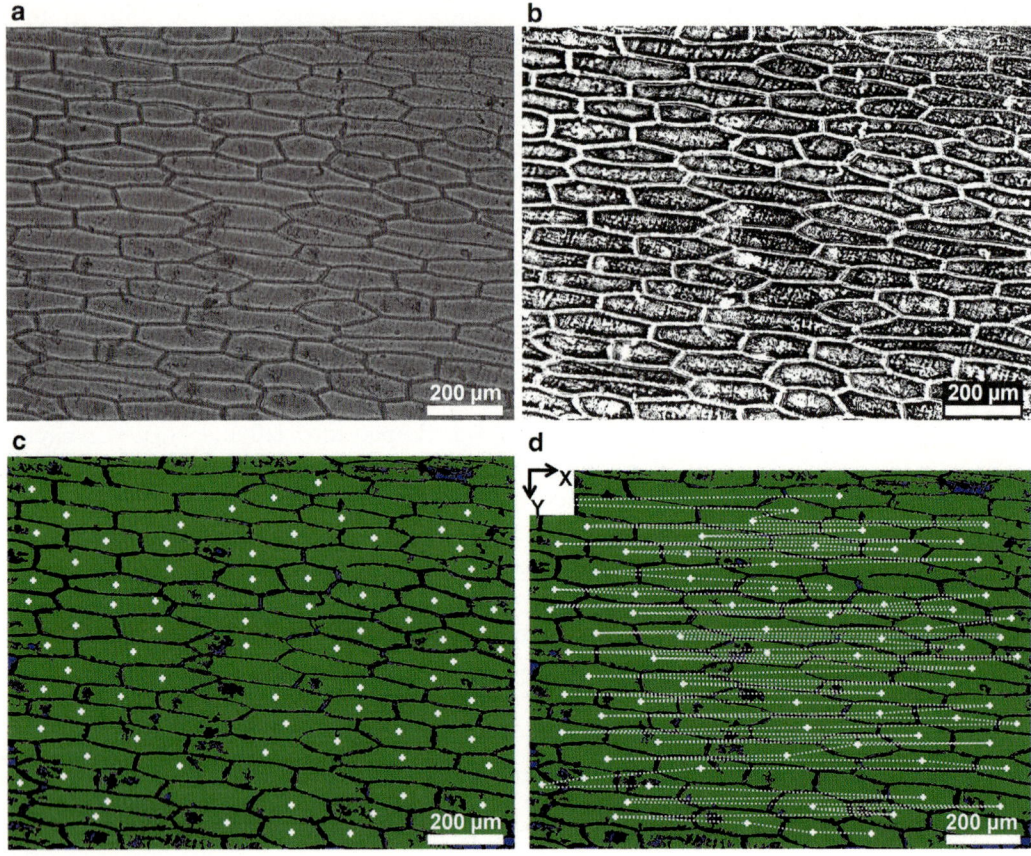

Fig. 1 Coordinate acquisition for single-cell ablation in an optical microscope image of *L. longiflorum* leaf adaxial epidermis. (**a**) Microscope image of *L. longiflorum* leaf adaxial epidermis is taken. (**b**) Cell walls are accentuated by thresholding the grayscale levels. (**c**) The image is binarized and the cells are recognized as objects. Centroids of cells (marked by *plus* signs) are determined. (**d**) Scanning path for the translation-stage movement exposing cell after cell to the ablation fiber is shown by *dashed line*. Origin is at the *top left corner* of the image

3.3 Translation-Stage Automation

1. The three-axis translation stage was configured and initialized through the stage controller.

2. The home position in the x–y plane and the elevation of the fiber tip in the z direction were optimized according to the sample geometry. There were two specific values for the fiber tip elevation. The "operational height" was determined by optimizing the laser ablation efficiency. The "relocation height" was more elevated to ensure unobstructed stage movement from one cell to another during scanning.

3. A set dwell time was determined based on the number of laser pulses needed to ablate a single cell.

3.4 LAESI-MS on Single Cells

1. The microscope slide holding the cell layer (e.g., *L. longiflorum* leaf abaxial epidermis or *A. cepa* epidermis) was carefully moved from the upright microscope to the translation stage in front of the mass spectrometer (*see* **Note 5**).

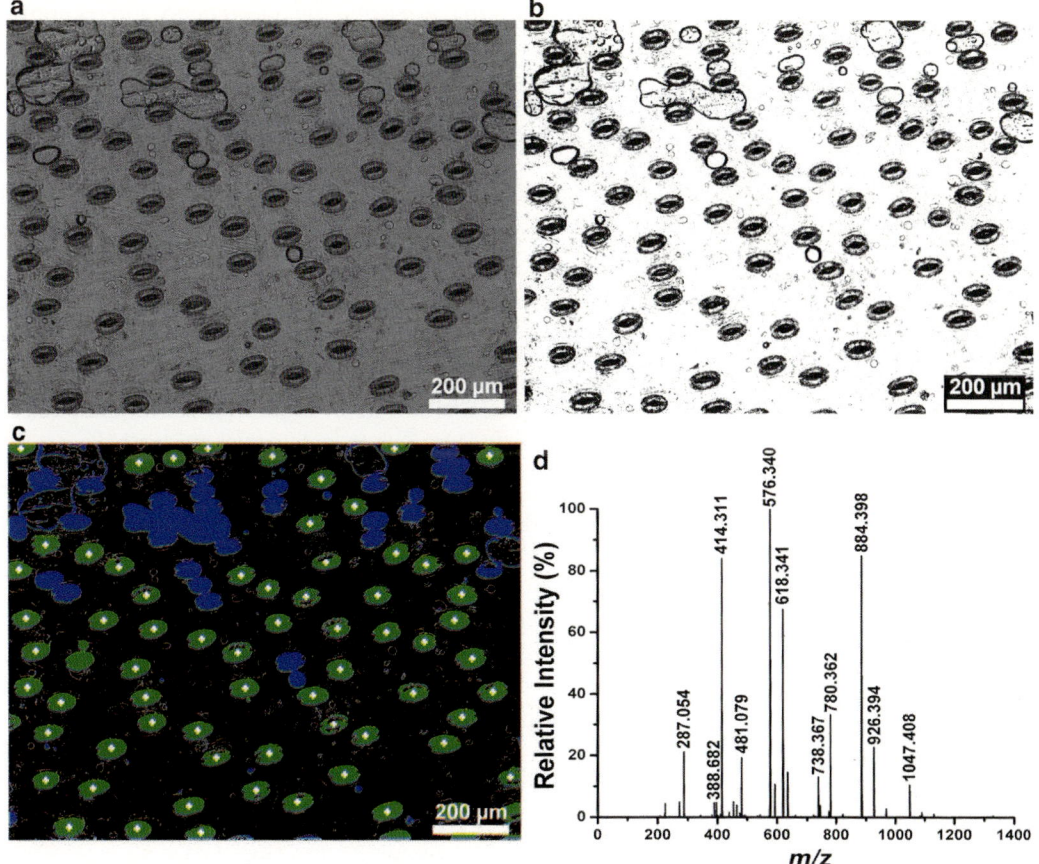

Fig. 2 (**a**) Microscope image of *L. longiflorum* leaf abaxial epidermis. (**b**) Thresholded image showing cell edge detection. (**c**) Centroids of guard cells with distinctive morphology are determined in the binarized image (marked by *plus* signs). (**d**) Mass spectrum of a pair of guard cells

2. The top-view and side-view cameras were adjusted to allow clear observation of the selected tissue and the etched fiber tip.

3. To accurately locate the origin of the coordinate system selected under the imaging microscope and position it under the fiber tip, the unique cell or feature in its environment was found through the top-view camera (*see* **Note 6**).

4. The fiber tip was lowered to a distance of 30–15 μm above the sample surface ("operational height") and adjusted in the *x–y* plane to be located above the origin of the coordinate system defined over the sample (*see* **Note 7**).

5. The syringe pump operating at a flow rate of 300 nL/min supplied the electrospray solution to the tapered stainless steel emitter. Stable electrospray was generated by applying a high voltage (2,800–3,000 V) on the metal union of the electrospray system.

6. The cell coordinates and scanning parameters were imported into the stage-control program.

7. The mid-IR laser was initialized and the pulse energy and repetition rate were optimized to enable the highest signal-to-noise ratio in a single-cell mass spectrum without affecting neighboring cells or breaking the fiber tip.

8. Acquisition parameters for the mass spectrometer (e.g., mass range: m/z 20–1,500, scan rate: 1 s/scan, positive ion acquisition mode) were selected.

9. The acquisition of mass spectra was initiated.

10. The pulses from mid-IR laser were fired at the first cell and simultaneously the stage-control program was started. The translation stage was directed to present the selected cells one by one to the tip of the ablation fiber. Figure 1d shows the path of movement on the *L. longiflorum* leaf adaxial epidermis.

11. When the data acquisition process was completed, all instrument components, including sample scanning, mid-IR laser, and electrospray, were stopped.

3.5 Cell-by-Cell Images

1. Mass spectra were analyzed to evaluate metabolites levels of individual cells. For example, in an experiment to study the molecular composition of cells with a particular morphology, the mass spectrum from a pair of guard cells of the *L. longiflorum* leaf abaxial epidermis was acquired (*see* Fig. 2d). Ions of interest, such as m/z 884.398 and 926.394, were identified as steroidal glycosides by separate tandem MS experiments and were consistent with previous studies [16, 17].

2. For cell-by-cell imaging, the peak intensities of selected ions were traced in ion chromatograms as individual cells were interrogated. The scan numbers in the chromatogram were correlated to the cells in the microscope image.

3. The ion intensities from each ablated cell were measured to construct a false color cell-by-cell molecular image reflecting the metabolite distribution in the tissue. Figure 3c, d shows the cell-by-cell molecular images for selected ions in a monolayer of *A. cepa* epidermis. Mass spectra for nonpigmented and purple cells show significant differences (*see* Fig. 3b for the comparison). The strong correlation of the pigment distribution, e.g., cyanidin malonyl glucoside with m/z 535.113 (*see* Fig. 3c), with the coloration of these cells in the optical image (*see* Fig. 3a), validates this cell-by-cell imaging approach. In contrast, a trisaccharide distribution with m/z 543.159 does not follow the coloration pattern (*see* Fig. 3d).

4   Notes

1. Direct contact with the high voltage applied to the electrospray emitter can cause electric shock that may result in severe injuries or death. Exposed electrical components were carefully

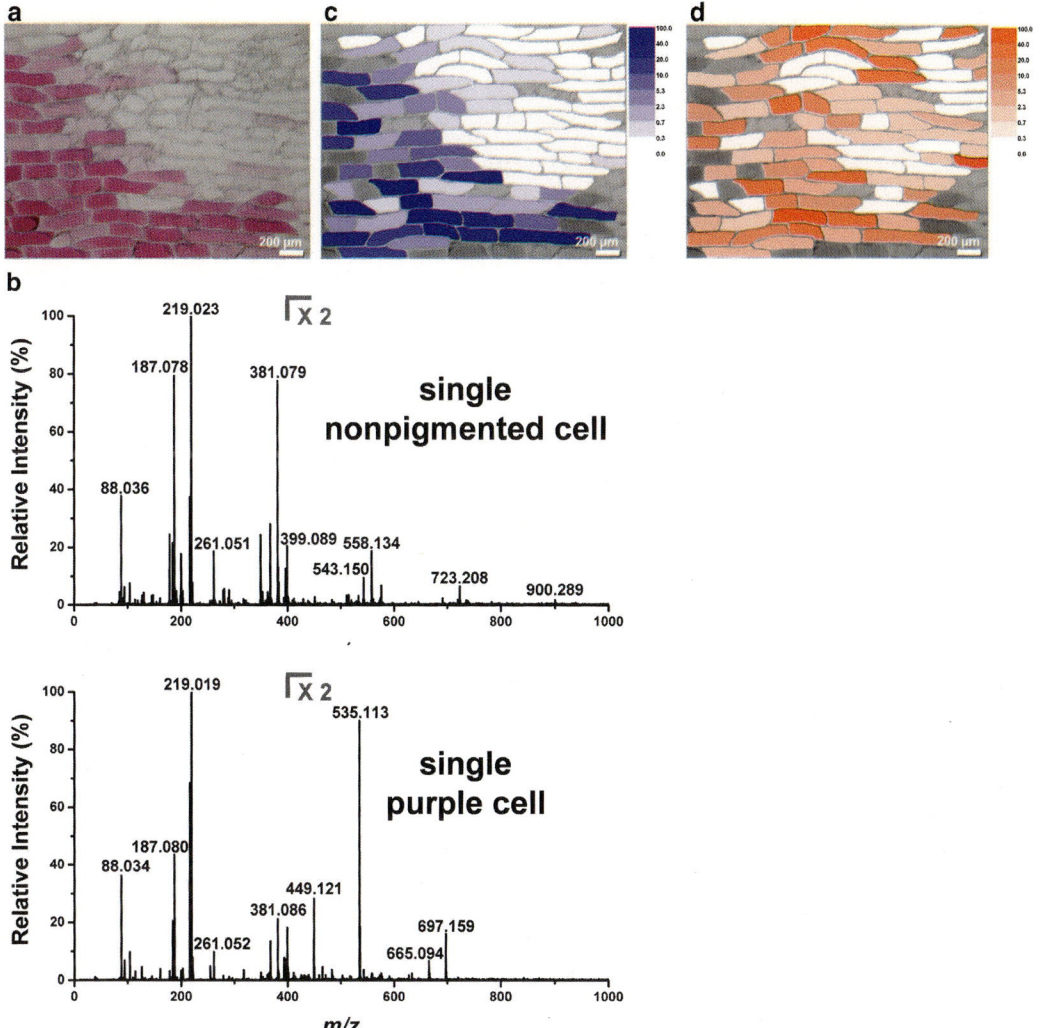

Fig. 3 (**a**) Microscope image of *A. cepa* bulb epidermal tissue showing nonpigmented and purple cells. (**b**) Mass spectra from (*top panel*) a single nonpigmented cell and (*bottom panel*) a single purple cell. (**c**) Cell-by-cell molecular image of cyanidin malonyl glucoside with *m/z* 535.113 is consistent with the distribution of purple color in the optical image (*see* Fig. 3a). (**d**) Cell-by-cell molecular image of trisaccharide with *m/z* 543.159 shows more uniform distribution

   shielded and appropriate signs were posted. It was forbidden to approach the high-voltage area during experiments.

2. Appropriate protection from laser beam exposure is necessary, including wearing mid-IR-range laser goggles and clothes with long sleeves.

3. The imaging methods of the upright microscope included bright-field and transmission illumination, and differential interference contrast mode. Images were obtained with high definition of the cell walls to facilitate automatic cell recognition.

4. Integrated morphometry measurements were performed on well-binarized images with clearly differentiated cells as objects. Filter parameters in the image analysis included object area, shape, and orientation to improve the measurement accuracy.

5. The sample slide mounted on the translation stage had to be placed horizontally flat in the x–y plane to maintain a constant distance between the fiber tip and the tissue surface over the studied area, and prevent the fiber tip from touching the sample during scanning.

6. The origin defined in the imaging microscope had to be found in the top-view observation microscope visualizing the single-cell LAESI experiment. The positioning of images under these two visualization systems had to be consistent and follow the same orientation.

7. Fine control of the distance (~25 μm) between the fiber tip and the sample surface was required for efficient ablation. Touching the sample surface by the etched fiber should be avoided to prevent damage to the cell and/or the fiber tip.

Acknowledgments

The authors acknowledge the financial support from the US National Science Foundation (Grant No. CHE-1152302) and the George Washington University Selective Excellence Fund. The GeO$_2$-based glass fibers were generously provided by Infrared Fiber Systems (Silver Spring, MD).

References

1. Chaurand P, Schwartz SA, Billheimer D, Xu BJ, Crecelius A, Caprioli RM (2004) Integrating histology and imaging mass spectrometry. Anal Chem 76:1145–1155

2. Miura D, Fujimura Y, Wariishi H (2012) In situ metabolomic mass spectrometry imaging: recent advances and difficulties. J Proteomics 75:5052–5060

3. Watrous JD, Dorrestein PC (2011) Imaging mass spectrometry in microbiology. Nat Rev Microbiol 9:683–694

4. Watrous JD, Alexandrov T, Dorrestein PC (2011) The evolving field of imaging mass spectrometry and its impact on future biological research. J Mass Spectrom 46:209–222

5. Goto-Inoue N, Hayasaka T, Zaima N, Setou M (2011) Imaging mass spectrometry for lipidomics. Biochim Biophys Acta 1811:961–969

6. Walch A, Rauser S, Deininger S-O, Höfler H (2008) MALDI imaging mass spectrometry for direct tissue analysis: a new frontier for molecular histology. Histochem Cell Biol 130:421–434

7. Wiseman JM, Ifa DR, Song Q, Cooks RG (2006) Tissue imaging at atmospheric pressure using desorption electrospray ionization (DESI) mass spectrometry. Angew Chem Int Ed 45:7188–7192

8. Nemes P, Barton AA, Li Y, Vertes A (2008) Ambient molecular imaging and depth profiling of live tissue by infrared laser ablation electrospray ionization mass spectrometry. Anal Chem 80:4575–4582

9. Nemes P, Barton AA, Vertes A (2009) Three-dimensional imaging of metabolites in tissues under ambient conditions by laser ablation electrospray ionization mass spectrometry. Anal Chem 81:6668–6675

10. Schwamborn K, Caprioli RM (2010) MALDI imaging mass spectrometry: painting molecular pictures. Mol Oncol 4:529–538

11. Nemes P, Woods AS, Vertes A (2010) Simultaneous imaging of small metabolites and lipids in rat brain tissues at atmospheric pressure by laser ablation electrospray ionization mass spectrometry. Anal Chem 82:982–988

12. Altschuler SJ, Wu LF (2010) Cellular heterogeneity: Do differences make a difference? Cell 141:559–563

13. Shrestha B, Vertes A (2009) In situ metabolic profiling of single cells by laser ablation electrospray ionization mass spectrometry. Anal Chem 81:8265–8271

14. Shrestha B, Nemes P, Vertes A (2010) Ablation and analysis of small cell populations and single cells by consecutive laser pulses. Appl Phys A 101:121–126

15. Shrestha B, Patt JM, Vertes A (2011) In situ cell-by-cell imaging and analysis of small cell populations by mass spectrometry. Anal Chem 83:2947–2955

16. Munafo JP, Gianfagna TJ (2011) Quantitative analysis of steroidal glycosides in different organs of Easter lily (Lilium longiflorum Thunb.) by LC-MS/MS. J Agric Food Chem 59:995–1004

17. Munafo JP, Ramanathan A, Jimenez LS, Gianfagna TJ (2010) Isolation and structural determination of steroidal glycosides from the bulbs of Easter lily (Lilium longiflorum Thunb.). J Agric Food Chem 58:8806–8813

Chapter 13

Laser Ablation Sample Transfer for Mass Spectrometry Imaging

Sung-Gun Park and Kermit K. Murray

Abstract

Infrared laser ablation sample transfer (IR-LAST) is a novel ambient sampling technique for mass spectrometry. In this technique, a pulsed mid-IR laser is used to ablate materials that are collected for mass spectrometry analysis; the material can be a solid sample or deposited on a sample target. After collection, the sample can be further separated or analyzed directly by mass spectrometry. For IR-LAST sample transfer tissue imaging using MALDI mass spectrometry, a tissue section is placed on a sample slide and material transferred to a target slide by scanning the tissue sample under a focused laser beam using transmission-mode (back side) IR laser ablation. After transfer, the target slide is analyzed using MALDI imaging. The spatial resolution is approximately 400 μm and limited by the spread of the laser desorption plume. IR-LAST for MALDI imaging provides several new capabilities including ambient sampling, area to spot concentration of ablated material, multiple ablation and analysis from a single section, and direct deposition on matrix-free nanostructured targets.

Key words Ambient sampling, Ambient mass spectrometry, Laser ablation, MALDI, MALDI imaging, NALDI

1 Introduction

Imaging mass spectrometry (IMS) is a form of mass spectrometry in which individual spectra are obtained at regular intervals over a sample surface, and the spatial distribution and relative abundance of molecules in the sample are obtained from the mass spectra and displayed as two-dimensional heat maps [1–8]. IMS is an effective tool not only for monitoring the spatial distribution of biomolecules but also for obtaining quantitative information on both known and unknown molecules from various samples. Such information can be obtained in a single experiment without chemical staining or target-specific reagents such as antibodies. MSI has broad applications in materials analysis, molecular biology, drug discovery, disease diagnosis, and therapy assessment by analyzing

Lin He (ed.), *Mass Spectrometry Imaging of Small Molecules*, Methods in Molecular Biology, vol. 1203, DOI 10.1007/978-1-4939-1357-2_13, © Springer Science+Business Media New York 2015

analytes, ranging from small to large molecules, in cells, tissues, organs, and even whole-body sections.

The most widely used IMS techniques are secondary ion mass spectrometry (SIMS) and matrix-assisted laser desorption (MALDI) [1, 7]. In SIMS, a focused primary ion beam impinges on a sample surface to produce secondary ions that are analyzed by MS. SIMS can produce images with spatial resolution as low as a few hundred nanometers but is limited to compounds with relatively low molecular weights due to high energy of the primary ion beam. In MALDI, a pulsed laser beam is used to desorb and ionize materials in a solid sample aided by a co-crystallized matrix. The mass range of MALDI is sufficient to detect molecules as large as proteins with a spatial resolution on the order of tens of micrometers. Both techniques, however, require the sample to be confined to the high-vacuum region of the instrument, and the addition of a matrix can lead to spatial dislocation of biomolecules on the sample surface.

In recent years, MS methods with direct sampling and ionization under ambient conditions have been introduced and used for IMS [6–9]. A variety of ambient mass spectrometry techniques have been introduced and used for analysis of molecules such as peptides, proteins, lipids, and metabolites. Desorption electrospray ionization (DESI) uses a spray of charged droplets from an electrospray source directed at the sample to extract analyte and form ions through solvent evaporation from charged secondary droplets [10]. For imaging, the spray is scanned across the sample surface while mass spectra are recorded. DESI MS imaging has been demonstrated for peptides, lipids, and drugs in animal and plant tissue sections with a spatial resolution of 100–500 μm [11–22].

Ambient sampling from the surface can be also achieved via direct liquid extraction, for example, with a liquid microjunction probe [23–25]. The probe consists of two concentric capillaries that come into near contact with a sample. The outer capillary supplies fresh extraction solvent to the sample surface and the inner capillary withdraws the extracted material and delivers it to an electrospray ion source. A spatial resolution of 500 μm has been achieved with this system [26]. A surface sampling system called nano-DESI uses separate extraction and sampling capillaries and has been used for imaging at a spatial resolution below 20 μm [27].

There are several laser-based approaches to ambient mass spectrometry [28]. In these techniques, the laser is used to remove material from a sample by desorption of molecules or ablation of particles. The removed material is then either ionized directly or delivered to a charge source such as an electrospray droplet or reagent ions. The laser-based sampling techniques typically allow better spatial resolution due to smaller size of the focused laser for resolution in the range of 10–100 μm [29]. Atmospheric pressure MALDI (AP-MALDI) uses a pulsed laser at atmospheric pressure to produce ions that are sampled into a mass spectrometer using a modified electrospray source. Imaging AP-MALDI has been demonstrated at a spatial resolution of 5 μm [30]. However, the application of AP-MALDI has been limited due to its low ionization efficiency at ambient conditions [31, 32].

Laser sample transfer to an electrospray source has been reported for both small and large biological molecules both with and without traditional MALDI matrices [33, 34]. These methods are known by various acronyms, but all use a pulsed laser to desorb or ablate material; ions are formed when the plume of laser-desorbed and -ablated material interacts with the electrospray. Electrospray-assisted laser desorption ionization (ELDI) uses a 337 nm UV laser and no matrix [33]. Matrix-assisted laser desorption electrospray ionization (MALDESI) uses a 337 nm UV laser and a MALDI matrix to aid the desorption process [34]. Infrared MALDESI [35, 36] and laser ablation electrospray ionization (LAESI) [37] use mid-infrared lasers at a wavelength near 3 μm for ambient ablation with electrospray ionization [38]. The IR laser electrospray approaches have been demonstrated for imaging with 100 μm [39] and 45 μm [40] spatial resolution.

Chemical ionization and discharge sources can be used for laser desorption/ablation ambient mass spectrometry, for example, laser ablation coupled to flowing atmospheric pressure afterglow (LA-FAPA) [41] and infrared laser ablation metastable-induced chemical ionization (IR-LAMICI) [42]. In LA-FAPA, the material is ablated by a 266 nm UV laser and transferred to a flowing afterglow of a helium in an atmospheric pressure glow discharge ionization source and ionized. In IR-LAMICI, IR-ablated materials from the sample surface interact with metastable ions from a discharge ion source and are ionized within the plume by chemical ionization. The IR LAMICI and LA-FAPA techniques were demonstrated with a spatial resolution of 300 μm [42] and 20 μm [41], respectively. Laser ablation can also be coupled with an inductively coupled plasma for the detection of metals in tissue [43].

Materials removed from samples under ambient conditions can be captured and later analyzed. For example, materials removed by DESI were collected and then analyzed by mass spectrometry, gas chromatography–mass spectrometry, and absorption spectroscopy [44, 45]. Laser-ablated materials can also be captured and analyzed with MALDI and ESI in a technique known as laser ablation sample transfer (LAST) [46–49]. For electrospray [46], biomolecules are ablated and captured in a solvent flow and the compounds separated by liquid chromatography [48] or capillary electrophoresis [50]. For MALDI, the material can be collected in a droplet for off-line analysis or transferred from a tissue section to a target for imaging.

In this chapter, we describe the protocol for sampling preparation using LAST for MALDI tissue imaging. An outline of the procedure is described in Subheading 3.

2 Materials

Prepare all solutions using ultrapure water and analytical grade reagents.

2.1 Reagents and Sample

1. DHB (matrix) solution: Dissolve 35 mg/mL of 2,5-dihydroxybenzoic acid (DHB) in a 1:1 (v/v) mixture of methanol and 0.1 % aqueous trifluoroacetic acid (TFA).

2. SA (matrix) solution: Dissolve 20 mg/mL of sinapinic acid (SA) in a 3:2 (v/v) mixture of acetonitrile (ACN) and 0.2 % aqueous TFA.

3. CHCA (matrix) solution: Dissolve 50 mg/mL of α-cyano-4-hydroxycinnamic acid (CHCA) in a 1:1 (v/v) mixture of methanol and 0.1 % aqueous TFA.

4. Nitrocellulose: Dissolve 5 mg/mL of nitrocellulose in a 4:1 (v/v) mixture of methanol and acetone.

5. Tissue sample: Store at −80 °C until it is sliced into 10 μm thick sections using a cryostat.

6. Peptide calibration standard: Dissolve a peptide mixture solution in a 125 μL mixture solution of ACN and 0.1 % trifluoroacetic acid (TFA) in a volume ratio of 1:2.

2.2 Mid-IR Ablation Sample Transfer

1. Indium tin oxide (ITO)-coated microscope slide.

2. Nanostructure-assisted laser desorption ionization (NALDI) or similar nanostructured target.

3. Thin-layer chromatography (TLC) sprayer.

4. Wavelength tunable pulsed infrared optical parametric oscillator.

5. Gold-coated mirrors for mid-IR light.

6. Plano-convex CaF$_2$ focusing lens (f = 50 mm).

7. Laser burn paper to measure laser spot size.

8. Three-axis computer-controlled motion stage.

2.3 Molecular Imaging and Data Analysis

1. MALDI mass spectrometer.

2. Slide adapter for mounting ITO coating slides for insertion into the MS instrument for MALDI analysis.

3. MALDI Imaging software.

3 Methods

Figure 1 shows the LAST system and MALDI images obtained from mouse brain tissue transferred to a target slide [51]. In this technique, the IR laser is used to transfer biomolecules from a

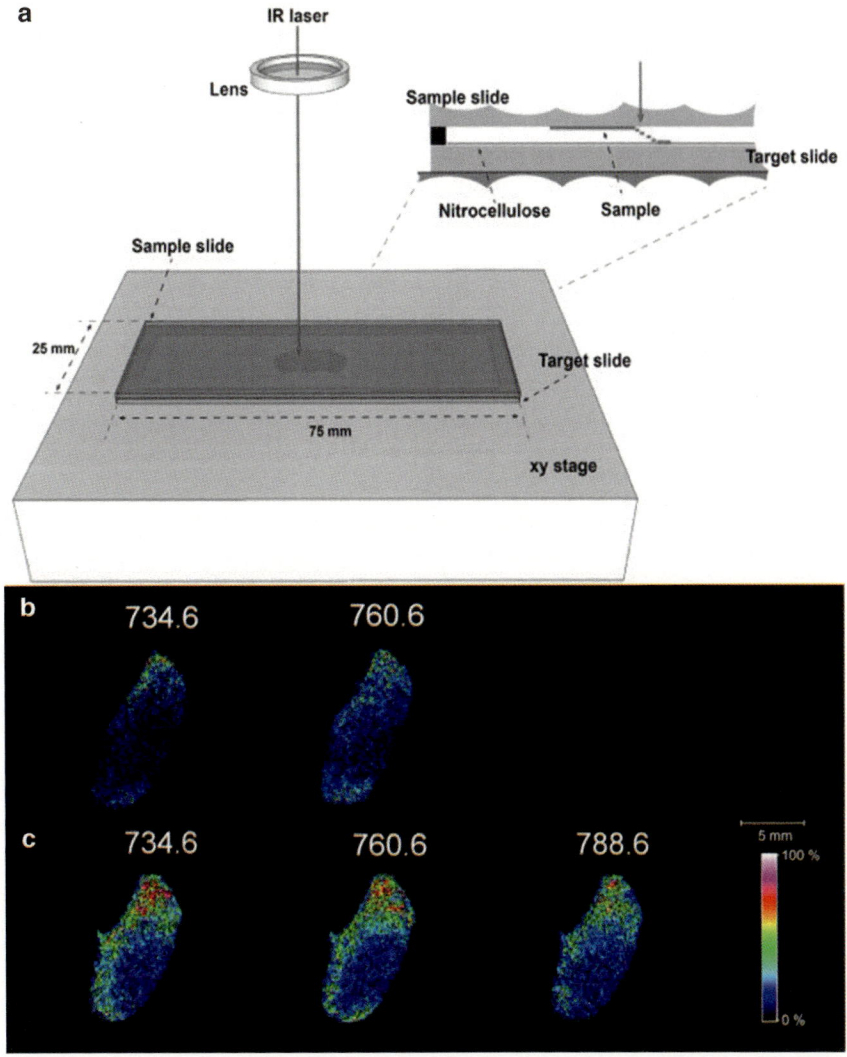

Fig. 1 (**a**) Schematic of the laser ablation sample transfer system for MALDI imaging. MALDI images of mouse brain sections prepared by IR laser ablation transfer to (**b**) matrix-pre-coated target and (**c**) target with subsequent matrix addition (reprinted with permission from Reference 51; Copyright 2012 American Chemical Society)

tissue section on a microscope slide (sample slide) for collection on a target slide or metal target for MALDI imaging. The target slide can be an ITO-coated microscope slide, a MALDI, or a nanostructured (NALDI) target. The ITO slide or MALDI target is coated with a thin layer of nitrocellulose from a solution of nitrocellulose using a TLC sprayer. For matrix-pre-coated targets, either DHB or SA matrix is sprayed onto the MALDI target on the nitrocellulose layer. For laser ablation transfer, a 10 μm thick tissue sample is mounted on a sample slide that is placed face down against the target. The gap between the slide and target is adjusted between 50 and 500 μm using different thicknesses of adhesive tape.

The slide and target are mounted on a computer-driven XY stage and the tissue is moved in two dimensions under the IR laser beam to transfer material from the sample slide to the target.

After the material is transferred from the sample slide to the slide, NALDI, or MALDI target, the latter is mounded in a slide adaptor and analyzed using MALDI mass spectrometry. Images are constructed from the mass spectra from using imaging software. For multiple images from a single sample slide using different target slides, a target slide is replaced with a new target slide after scanning the tissue with the IR laser. The sample and new target slides are mounted on the XY stage and scanned again under the IR laser. For concentrating ablated material from large area of tissue onto a single spot on the target slide, the sample slide is moved under the laser beam and the target slide is held in place to capture the material in one spot.

To demonstrate the spatial resolution that can be obtained using laser ablation sample transfer, a peptide mass standard was laser ablation transferred from a sample target deposit to a set of lines on the target slide and imaged. For this experiment, a 1 mM solution of the peptide angiotensin II was spray deposited onto a microscope slide. After air-drying, a glycerol solution was sprayed on the deposit to wet it. The slide was placed in the target with a 70 μm spacing and irradiated at 3 kJ/m² laser fluence. The gaps between laser-ablated lines were 1 mm, 800 μm, 600 μm, 400 μm, and 300 μm. From this image, it can be seen that the minimum distinguishable spacing of the transferred lines on the target is 400 μm.

3.1 Preparation of Sample Slides

1. Transfer a mouse brain onto a cooled sample stage (−20 °C) of a cryostat.

2. Obtain a mouse brain section of 10 μm thickness.

3. Thaw-mount the tissue on a conductive side of ITO-coated microscope slide (sample slide). A multimeter can be used to determine which side is conductive.

4. Store the tissue sample slides at −80 °C before using.

3.2 Preparation of Target Slides

Before preparation of a target slide, optimize the time for spraying and drying, and the distance between the nozzle and the sample to make a homogenous coating on the target (*see* **Notes 1–3**).

1. Use a TLC nebulizer with nitrogen gas to spray the nitrocellulose or matrix solution.

2. Mount an ITO-coated microscope slide perpendicularly in a hood.

3. Add the prepared nitrocellulose solution into the TLC sprayer.

4. Set the distance between the slide and sprayer to about 30 cm.

5. Spray the ITO glass slide for 20 s and dry for 2 min.

6. Repeat **step 5** nine more times (ten times in all).

7. For a matrix-pre-coated target, spray a matrix solution onto the nitrocellulose-coated slide for 20 s and dry for 2 min.

8. Repeat **step 7** nine more times (ten times in all).

3.3 Laser Ablation Sample Transfer for MALDI Imaging

1. Optimize the gap between the sample and target slides and optimize the laser pulse energy. To optimize the gap and laser energy, a 1 mM solution of angiotensin II is sprayed on a sample slide. After air-drying, a glycerol solution is sprayed on the sample slide to assist the IR laser ablation. The sample slide faces toward the target slide, and the gap between the slide and target is adjusted using different thicknesses of adhesive tape. The slides with different gaps are irradiated at a single spot at different laser energies. After transfer of material from the sample slide, the target slides are spray coated with CHCA matrix and imaged using MALDI. For example, Fig. 2 shows MALDI images of peptide spots transferred by IR laser ablation at different IR laser fluences and different distances between the sample and target slide. At a spacing of 70 μm and 3 kJ/m^2 laser fluence, a small peptide spot size is observed although less material is transferred (*see* **Note 4**).

2. Place sample slide face down against the target slide with a spacing of 70 μm.

3. Mount the two slides on the translation stage.

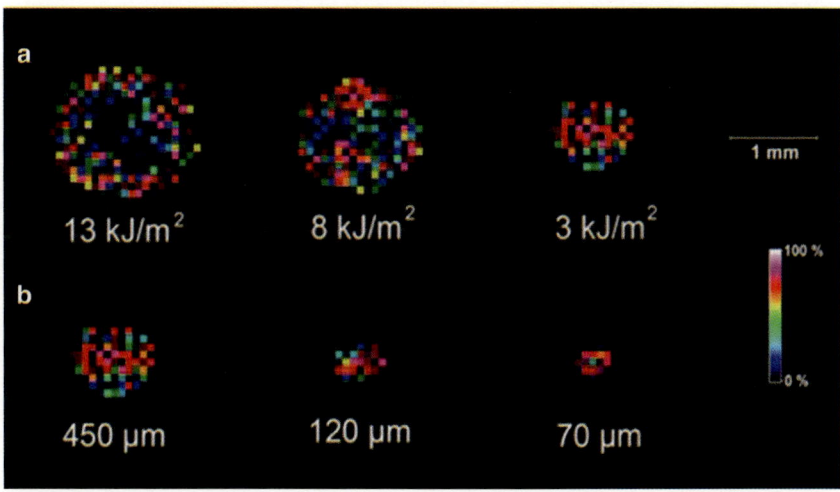

Fig. 2 MALDI image of peptide spots transferred by IR laser ablation (**a**) at the indicated laser fluences and at a spacing of 450 μm and (**b**) at the indicated distances and at a laser fluence of 3 kJ/m^2 (reprinted with permission from Reference 51; Copyright 2012 American Chemical Society)

4. Use gold-coated first-surface mirrors and CaF_2 lens to focus the mid-IR laser beam onto the sample at normal incidence in transmission mode (backside irradiation).

5. Scan the sample and target slides under the IR laser beam to transfer material from the sample to target slides. The linear velocity of the stage is 30 μm/s, and a serpentine pattern with 20 μm raster line spacing is traced. The laser wavelength is 3 μm and the number of laser shot is 200 at 20 Hz repetition rate (*see* **Note 5**).

6. Remove the target slide and mount it in the adaptor. If the target slide is not pre-coated with MALDI matrix (target slide coated with nitrocellulose), apply matrix to the transferred material using the TLC sprayer as described in Subheading 3.2 (**steps 7** and **8**) before MALDI imaging.

3.4 Multiple Tissue Transfers

1. After first ablating a tissue sample to transfer material to a target slide as described in Subheading 3.3, the target slide is replaced with a new target slide (coated with matrix or a NALDI target).

2. Place the sample side with the first scanned tissue sample on the new target slide, scan again under the IR laser beam, and run MALDI from the target slide.

3.5 Area to Spot Transfer

1. Mount a sample side on translation stage.

2. Hold a target slide in place (*see* **Note 6**).

3. Set the distance between the sample side and the target slide to 100 μm using the z-axis (*see* **Note 7**).

4. Scan the sample under the IR laser beam to transfer material from an area of the tissue to a spot on the target slide.

5. Add 2 μL of DHB matrix to the target and perform MALDI.

3.6 MALDI Imaging from a Sample Slide

1. Mix equal volumes of a peptide standard solution with a CHCA matrix solution.

2. Apply 1 μL of the solution to an ITO glass slide and dry at room temperature for MS.

3. Mount a sample slide and the peptide standard slide in slide adapter.

4. Insert the slide adapter into a MALDI mass spectrometer.

5. Perform mass calibration using the peptide standard.

6. After mass calibration, start the MS imaging run to scan the tissue sample and collect MALDI mass spectra.

7. After collecting mass spectra, reconstruct MALDI images.

4   Notes

1. The spray-coating step creates a uniform nitrocellulose/matrix layer on the surface. Smaller distances between the sprayer and the target slide and longer spraying times can result in over-wetting the slide and an inhomogeneous layer.

2. The water content in the tissue sample is a critical factor for mid-IR laser ablation sample transfer. Therefore, it is important to prevent the tissue from drying.

3. Before mounting the slides on the translation stage for material transfer from sample slide to target slide, the area of the tissue sample should be marked on the back of the target slide around the tissue area. The mark is useful for finding the transferred tissue area when running MALDI imaging.

4. At greater distances between the slides, the size of the spot of transferred material is larger due to the radial dispersion of the plume of ablated material. At high laser fluencies, the spot size of the transferred material on the MALDI target is much larger than the IR laser spot size. In addition, a donut-shaped image is observed due to ablation of the transferred sample or due to hydrodynamic ejection of material, which can result in removal of material from the sides of the ablation crater [52].

5. A large number of laser shots in the first laser ablation transfer can lead to low material transfer for the second image.

6. Before mounting the target slide, mark the spot on the backside of a target at the point that the ablated material will be transferred. In this experiment, the transferred area on the target slide is small (less than 120 µm), and it can be difficult to locate the spot of transferred material. The marked spot on the target indicates the transferred material.

7. In an area to spot transfer, a smaller distance between the sides produces higher concentration of transferred material.

Acknowledgments

This work was supported by the National Science Foundation, Grant Number CHE-1152106.

References

1. McDonnell LA, Heeren RMA (2007) Imaging mass spectrometry. Mass Spectrom Rev 26:606–643

2. Murayama C, Kimura Y, Setou M (2009) Imaging mass spectrometry: principle and application. Biophys Rev 1:131–139

3. Boxer SG, Kraft ML, Weber PK (2009) Advances in imaging secondary ion mass spectrometry for biological samples. Annu Rev Biophys 38:53–74

4. Schwartz S, Caprioli R (2010) Imaging mass spectrometry: viewing the Future. In: Rubakhin

SS, Sweedler JV (eds) Mass spectrometry imaging, vol. 656. Humana, New York, pp 3–19

5. Schwamborn K, Caprioli RM (2010) MALDI imaging mass spectrometry—painting molecular pictures. Mol Oncol 4:529–538

6. Chughtai K, Heeren RMA (2010) Mass spectrometric imaging for biomedical tissue analysis. Chem Rev 110:3237–3277

7. Amstalden van Hove ER, Smith DF, Heeren RMA (2010) A concise review of mass spectrometry imaging. J Chromatogr 1217:3946–3954

8. Chaurand P (2012) Imaging mass spectrometry of thin tissue sections: a decade of collective efforts. J Proteomics 75:4883–4892

9. Wu C, Dill AL, Eberlin LS et al (2013) Mass spectrometry imaging under ambient conditions. Mass Spectrom Rev 32:218–243

10. Takáts Z, Wiseman JM, Gologan B et al (2004) Mass spectrometry sampling under ambient conditions with desorption electrospray ionization. Science 306:471–473

11. Dill AL, Ifa DR, Manicke NE et al (2009) Mass spectrometric imaging of lipids using desorption electrospray ionization. J Chromatogr B 877:2883–2889

12. Esquenazi E, Dorrestein PC, Gerwick WH (2009) Probing marine natural product defenses with DESI-imaging mass spectrometry. Proc Natl Acad Sci U S A 106:7269–7270

13. Ifa DR, Wiseman JM, Song Q et al (2007) Development of capabilities for imaging mass spectrometry under ambient conditions with desorption electrospray ionization (DESI). Int J Mass Spectrom 259:8–15

14. Kertesz V, Van Berkel GJ (2008) Improved imaging resolution in desorption electrospray ionization mass spectrometry. Rapid Commun Mass Spectrom 22:2639–2644

15. Kertesz V, Van Berkel GJ (2008) Scanning and surface alignment considerations in chemical imaging with desorption electrospray mass spectrometry. Anal Chem 80:1027–1032

16. Kertesz V, Van Berkel GJ, Vavrek M et al (2008) Comparison of drug distribution images from whole-body thin tissue sections obtained using desorption electrospray ionization tandem mass spectrometry and autoradiography. Anal Chem 80:5168–5177

17. Lane AL, Nyadong L, Galhena AS et al (2009) Desorption electrospray ionization mass spectrometry reveals surface-mediated antifungal chemical defense of a tropical seaweed. Proc Natl Acad Sci U S A 106:7314–7319

18. Wiseman JM, Ifa DR, Zhu Y et al (2008) Desorption electrospray ionization mass spectrometry: imaging drugs and metabolites in tissues. Proc Natl Acad Sci U S A 105:18120–18125

19. Wu C, Ifa DR, Manicke NE et al (2009) Rapid, direct analysis of cholesterol by charge labeling in reactive desorption electrospray ionization. Anal Chem 81:7618–7624

20. Manicke NE, Nefliu M, Wu C et al (2009) Imaging of lipids in atheroma by desorption electrospray ionization mass spectrometry. Anal Chem 81:8702–8707

21. Eberlin LS, Ifa DR, Wu C et al (2010) Three-dimensional visualization of mouse brain by lipid analysis using ambient ionization mass spectrometry. Angew Chem Int Ed 49:873–876

22. Wiseman JM, Ifa DR, Song Q et al (2006) Tissue imaging at atmospheric pressure using desorption electrospray ionization (DESI) mass spectrometry. Angew Chem Int Ed 45:7188–7192

23. Van Berkel GJ, Sanchez AD, Quirke JME (2002) Thin-layer chromatography and electrospray mass spectrometry coupled using a surface sampling probe. Anal Chem 74:6216–6223

24. Kertesz V, Ford MJ, Van Berkel GJ (2005) Automation of a surface sampling probe/electrospray mass spectrometry system. Anal Chem 77:7183–7189

25. Kertesz V, Van Berkel GJ (2010) Liquid microjunction surface sampling coupled with high-pressure liquid chromatography—electrospray ionization-mass spectrometry for analysis of drugs and metabolites in whole-body thin sections. Anal Chem 82:5917–5921

26. Van Berkel GJ, Kertesz V, Koeplinger KA et al (2008) Liquid microjunction surface sampling probe electrospray mass spectrometry for detection of drugs and metabolites in thin tissue sections. J Mass Spectrom 43:500–508

27. Lanekoff I, Heath BS, Liyu A et al (2012) Automated platform for high-resolution tissue imaging using nanospray desorption electrospray ionization mass spectrometry. Anal Chem 84:8351–8356

28. Huang M-Z, Yuan C-H, Cheng S-C et al (2010) Ambient ionization mass spectrometry. Annu Rev Anal Chem 3:43–65

29. Norris JL, Caprioli RM (2013) Analysis of tissue specimens by matrix-assisted laser desorption/ionization imaging mass spectrometry in biological and clinical research. Chem Rev 113:2309–2342

30. Guenther S, Römpp A, Kummer W et al (2011) AP-MALDI imaging of neuropeptides in mouse pituitary gland with 5 μm spatial

resolution and high mass accuracy. Int J Mass Spectrom 305:228–237

31. Peng IX, Ogorzalek Loo RR, Margalith E et al (2010) Electrospray-assisted laser desorption ionization mass spectrometry (ELDI-MS) with an infrared laser for characterizing peptides and proteins. Analyst 135:767–772

32. Nemes P, Barton AA, Li Y et al (2008) Ambient molecular imaging and depth profiling of live tissue by infrared laser ablation electrospray ionization mass spectrometry. Anal Chem 80:4575–4582

33. Shiea J, Huang M-Z, Hsu H-J et al (2005) Electrospray-assisted laser desorption/ionization mass spectrometry for direct ambient analysis of solids. Rapid Commun Mass Spectrom 19:3701–3704

34. Sampson J, Hawkridge A, Muddiman D (2006) Generation and detection of multiply-charged peptides and proteins by matrix-assisted laser desorption electrospray ionization (MALDESI) Fourier transform ion cyclotron resonance mass spectrometry. J Am Soc Mass Spectrom 17:1712–1716

35. Sampson J, Murray K, Muddiman D (2009) Intact and top-down characterization of biomolecules and direct analysis using infrared matrix-assisted laser desorption electrospray ionization coupled to FT-ICR mass spectrometry. J Am Soc Mass Spectrom 20:667–673

36. Rezenom YH, Dong J, Murray KK (2008) Infrared laser-assisted desorption electrospray ionization mass spectrometry. Analyst 133:226–232

37. Nemes P, Vertes A (2007) Laser ablation electrospray ionization for atmospheric pressure, in vivo, and imaging mass spectrometry. Anal Chem 79:8098–8106

38. Murray KK (2006) Infrared MALDI. In: Caprioli RM, Gross ML (eds) Encyclopedia of mass spectrometry, vol 6. Elsevier, Amsterdam

39. Nemes P, Vertes A (2010) Atmospheric-pressure molecular imaging of biological tissues and biofilms by LAESI mass spectrometry. J Vis Exp 43:1–4

40. Robichaud G, Barry J, Garrard K et al (2013) Infrared matrix-assisted laser desorption electrospray ionization (IR-MALDESI) imaging source coupled to a FT-ICR mass spectrometer. J Am Soc Mass Spectrom 24:92–100

41. Shelley JT, Ray SJ, Hieftje GM (2008) Laser ablation coupled to a flowing atmospheric pressure afterglow for ambient mass spectral imaging. Anal Chem 80:8308–8313

42. Galhena AS, Harris GA, Nyadong L et al (2010) Small molecule ambient mass spectrometry imaging by infrared laser ablation metastable-induced chemical ionization. Anal Chem 82:2178–2181

43. Sabine Becker J (2013) Imaging of metals in biological tissue by laser ablation inductively coupled plasma mass spectrometry (LA–ICP–MS): state of the art and future developments. J Mass Spectrom 48:255–268

44. Kamali AS, Thompson JG, Bertman S et al (2011) Spray desorption collection of free fatty acids onto a solid phase microextraction fiber for trap grease analysis in biofuel production. Anal Method 3:683–687

45. Venter AR, Kamali A, Jain S et al (2010) Surface sampling by spray-desorption followed by collection for chemical analysis. Anal Chem 82:1674–1679

46. Ovchinnikova OS, Kertesz V, Van Berkel GJ (2011) Combining laser ablation/liquid phase collection surface sampling and high-performance liquid chromatography–electrospray ionization-mass spectrometry. Anal Chem 83:1874–1878

47. Park S-G, Murray K (2011) Infrared laser ablation sample transfer for MALDI and electrospray. J Am Soc Mass Spectrom 22:1352–1362

48. Park S-G, Murray KK (2012) Infrared laser ablation sample transfer for on-line liquid chromatography electrospray ionization mass spectrometry. J Mass Spectrom 47:1322–1326

49. Huang M-Z, Jhang S-S, Cheng C-N et al (2010) Effects of matrix, electrospray solution, and laser light on the desorption and ionization mechanisms in electrospray-assisted laser desorption ionization mass spectrometry. Analyst 135:759–766

50. Park S-G, Murray KK (2013) Ambient laser ablation sampling for capillary electrophoresis mass spectrometry. Rapid Commun Mass Spectrom 27:1673–1680

51. Park S-G, Murray KK (2012) Infrared laser ablation sample transfer for MALDI imaging. Anal Chem 84:3240–3245

52. Fan X, Murray KK (2009) Wavelength and time-resolved imaging of material ejection in infrared matrix-assisted laser desorption. J Phys Chem A 114:1492–1497

Chapter 14

Nanostructure Imaging Mass Spectrometry: The Role of Fluorocarbons in Metabolite Analysis and Yoctomole Level Sensitivity

Michael E. Kurczy, Trent R. Northen, Sunia A. Trauger, and Gary Siuzdak

Abstract

Nanostructure imaging mass spectrometry (NIMS) has become an effective technology for generating ions in the gas phase, providing high sensitivity and imaging capabilities for small molecules, metabolites, drugs, and drug metabolites. Specifically, laser desorption from the nanostructure surfaces results in efficient energy transfer, low background chemical noise, and the nondestructive release of analyte ions into the gas phase. The modification of nanostructured surfaces with fluorous compounds, either covalent or non-covalent, has played an important role in gaining high efficiency/sensitivity by facilitating analyte desorption from the nonadhesive surfaces, and minimizing the amount of laser energy required. In addition, the hydrophobic fluorinated nanostructure surfaces have aided in concentrating deposited samples into fine micrometer-sized spots, a feature that further facilitates efficient desorption/ionization. These fluorous nanostructured surfaces have opened up NIMS to very broad applications including enzyme activity assays and imaging, providing low background, efficient energy transfer, nondestructive analyte ion generation, super-hydrophobic surfaces, and ultra-high detection sensitivity.

Key words Nanostructure imaging mass spectrometry (NIMS), Desorption/ionization on silicon mass spectrometry (DIOS-MS), Metabolites, Mass spectrometry imaging

1 Introduction

Desorption mass spectrometry has undergone significant advancements since it was first developed more than a century ago [1]. A major improvement occurred in the early 1980s, with the development of matrix-assisted laser desorption/ionization (MALDI), a method of nondestructively transferring laser energy to the analyte by using a light-absorbing organic matrix [2, 3]. However, the use of organic matrices can present interference when attempting to detect small molecules less than 500 Da (e.g., metabolites). Therefore, in 1999 a matrix-free nanostructure imaging mass spectrometry (NIMS) strategy for mass spectrometry was introduced

Lin He (ed.), *Mass Spectrometry Imaging of Small Molecules*, Methods in Molecular Biology, vol. 1203,
DOI 10.1007/978-1-4939-1357-2_14, © Springer Science+Business Media New York 2015

based on using pulsed-laser desorption/ionization with a silicon nanostructured surface [4]. This method, originally called desorption/ionization on silicon mass spectrometry (DIOS-MS), uses laser irradiation to desorb and ionize analytes from a porous silicon surface, eliminating the need for organic matrices and thus extending the measurable mass below 500 Da [4]. Surface modifications of silicon nanostructured surfaces were later found to allow more efficient ion generation and resistance to oxidation [5, 6]. And more recently, the introduction of liquid fluorous compounds onto the nanostructured surface to form clathrates has resulted in improved detection capabilities as well as the ability to perform high-resolution imaging [7–10]. In this chapter, we discuss the possible mechanisms behind nanostructure desorption/ionization and the ultrahigh sensitivity that can be achieved with NIMS.

2   Nanostructure-Based Desorption/Ionization

One of the unique features of the NIMS desorption/ionization approach is its large surface area. High-surface-area porous silicon nanostructures facilitate efficient laser absorption and aid in the desorption/ionization of intact molecular ions through a laser-induced rearrangement of the surface structure [11–14] (Fig. 1). The large surface area (as large as 200 m^2/cm^2) can reduce the melting point of silicon; therefore laser-induced surface restructuring is thought to be the driver for analyte desorption [12]. The process is also highly dependent on the laser energy which directly correlates with ion generation. The low-threshold laser energy required for ion generation (10 mJ/cm^2), when compared to other desorption/ionization techniques like MALDI (40 mJ/cm^2), suggests that desorption/ionization is driven by surface restructuring and is not strictly a thermal process [12]. Similarly, the silicon nanowire [15], silicon nanopost arrays (NAPA) [16], laser-induced silicon microcolumn arrays (LISMA) [17], and other nanostructure-based techniques [14] likely work in a similar fashion; increased surface area typically lowers the laser energy required for analyte desorption.

Hydrophobic fluorous materials introduced into nanostructured surfaces have also played an important role in producing surfaces that allow for improved performance for NIMS including enhanced sensitivity (Fig. 2). Three different methods have been developed to incorporate fluorous compounds within porous silicon nanostructures. First, silicon nanostructures have been designed with a covalent pentafluorophenyl modification to reduce analyte adhesion and protect the porous surface from oxidation [5]. A second method has applied the addition of fluorous surfactants, such as perfluoroundecanoic acid, with the pentafluorophenyl-modified silicon surface. These surfaces have been shown to be more effective at reducing analyte adhesion and

Nanostructured surfaces

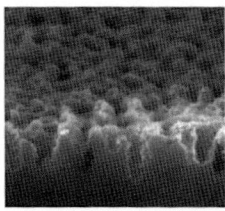 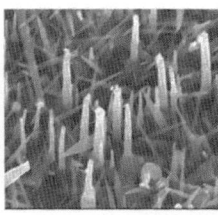 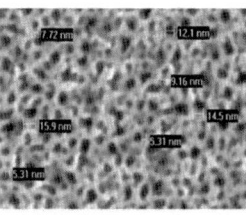

UV absorbing

thermal insulator

high surface area
(~200 m²/cm³)

Fig. 1 Electron micrographs of silicon-based nanostructure surfaces used in NIMS experiments. A unique feature of these surfaces is that they are UV-absorbing thermal insulators with a large surface area, facilitating their unique desorption/ionization properties

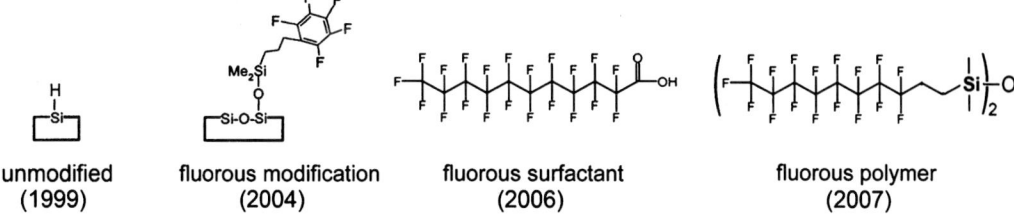

| unmodified | fluorous modification | fluorous surfactant | fluorous polymer |
| (1999) | (2004) | (2006) | (2007) |

Fig. 2 The evolution of fluorous modifications on the nanostructured surfaces, including unmodified surfaces in 1999 [4], chemical modification in 2004 [5], surfactants in 2006 [6], and teflon-like fluorous polymers such as bis(heptadecafluoro-1,1,2,2-tetrahydrodecyl)tetramethyl-disiloxane in 2007 [7]

improving desorption/ionization efficiency [6]. The third method, introduced in 2007, employs fluorous siloxanes as liquid initiators to coat the porous silicon nanostructure surface and further minimize analyte adhesion [7]. With this NIMS technology, it was found that fluorous siloxane initiators did not absorb laser light or ionize, and therefore do not contribute chemical noise in the spectrum, a very important aspect of the NIMS design. Subsequent laser-induced heating transfers energy to the trapped liquid phase, causing rapid initiator vaporization and desorption/ionization of the intact analytes without fragmentation. Among its features is that this surface is stable in ambient air, has an expanded mass range, and can be used to analyze biofluids and image tissues (Fig. 3). The versatility of the fluorinated NIMS platform has now been demonstrated for a large variety of analytes, ranging from metabolites and drugs to peptides and proteins [4–7].

3 Ultrahigh-Sensitivity Detection

The ultrahigh sensitivity that can be obtained with NIMS has been successfully demonstrated with specific analytes down to the yoctomole level as shown in Fig. 4. The first report of yoctomole sensitivity with NIMS was using a pentafluorophenyl-silylated

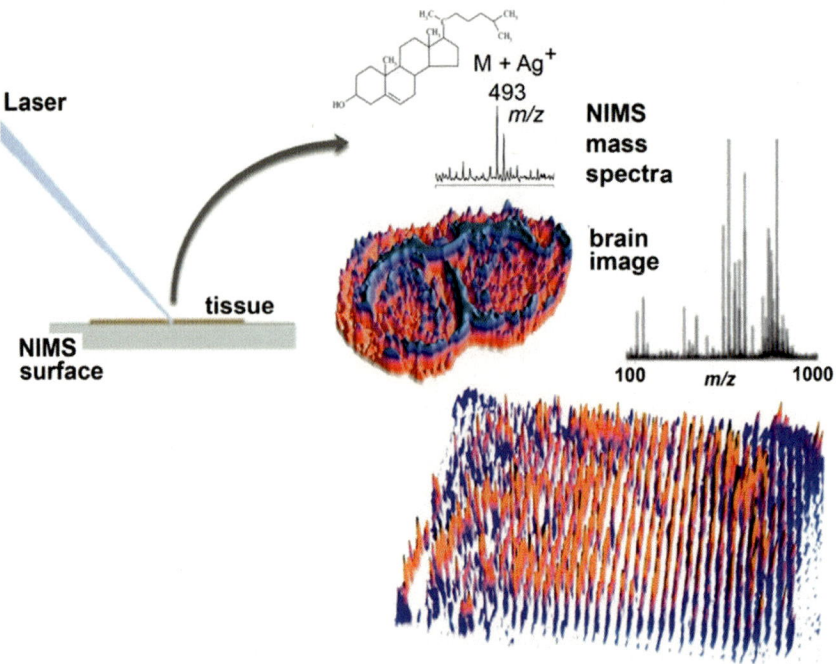

Fig. 3 Nanostructure imaging mass spectrometry (NIMS) of a brain tissue and also imaging of a plate containing 1,500 discrete chemical entities spotted on the NIMS surface

nanostructure silicon surface to analyze des-Arg9-bradykinin (des-Arg9-bradykinin is commonly used by instrument manufacturers to test sensitivity). Here a series of dilution experiments was carried out to ultimately demonstrate a lower limit of detection for the peptide at 480 molecules (800 ymol) (Fig. 3a) [5]. Similarly, NIMS was also found to have yoctomole detection for small molecules where lower limits of detection of 700 ymol for verapamil [18] and 650 ymol for propafenone have been observed [19] (Fig. 3b). Given the significance of this unprecedented sensitivity, the experiments were replicated on ten separate occasions by three different individuals.

4   Mechanistic Discussions

An important question to consider is why NIMS is inherently more sensitive than traditional matrix-assisted approaches such as MALDI, especially given that these experiments are performed with the same instrumentation. While very impressive, MALDI with high sensitivity (low zeptomole) has been achieved by Keller and Li [20], MALDI however is typically 50 times less sensitive than NIMS. To assess this difference in sensitivity, sample deposition was initially examined as this is a key feature that differs

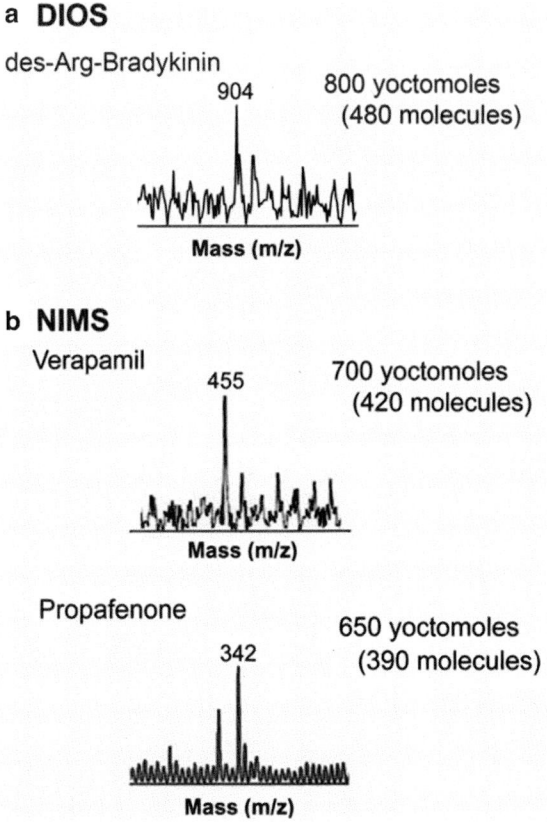

Fig. 4 High-sensitivity nanostructure imaging mass spectrometry (NIMS) experiments. Detection limit of (**a**) 480 molecules (800 ymol) for des-Arg9-bradykinin using pentafluorophenyl-functionalized porous silicon and (**b**) 420 molecules (700 ymol) for verapamil and 390 molecules (650 ymol) of propafenone using a bis(tridecafluoro-1,1,2,2-tetrahydrooctyl) tetramethyldisiloxane initiator

between NIMS and MALDI. In typical NIMS experiments the sample droplet is spotted directly onto the nanostructured surface. The unique nonadhesive surface properties of the fluorinated modifications and coatings used for NIMS not only reduce adhesion of the analyte facilitating desorption, but also the hydrophobic nature of the coating results in the formation of small aqueous droplets that concentrate the analyte on the surface. Simply put, the aqueous droplet being hydrophilic minimizes its contact area with the fluorinated coating and dries in a smaller spot concentrating the analyte. Another advantage of this technique is in its application to real biological samples and biofluids, which often contain salts and buffers which are detrimental to mass spectrometry. The process of analyte concentration on the hydrophobic fluorinated coating separates the salts to the outer edges, essentially "cleaning up" the analyte for analysis. The hydrophobic-hydrophobic interaction occurring between the fluorocarbon and the analyte serves to corral these molecules on the nanostructured surface, minimizing the number of analyte molecules in a given area necessary to produce the analyte signal. In many cases it is possible to adsorb analyte onto the fluorous surface directly from the sample droplet to minimize the effects of interferences within the sample (e.g., salts, proteins). This is again thought to be a result of the high surface energy at the fluorous-aqueous boundary that drives adsorption of molecules with amphipathic characteristics to the interface. The concentration effect can easily improve the detection sensitivity by a factor of 10–100. This can enable a signal to be generated from a small amount of material that is quickly consumed with a few laser shots.

Another distinguishing feature between nanostructure-based desorption/ionization and MALDI is that MALDI incorporates analytes into the matrix crystals which can affect its sensitivity, as does the ionization of the matrix materials, causing analyte signal suppression. Thus in MALDI, the spatial limitation of analytes exists both laterally across the surface as well as being dependent on the matrix crystal thickness/depth (microns to millimeters in size). The resulting laser-induced ablation following each laser shot introduces new crystal surfaces from which a signal can be produced. The crystal thickness allows for a continuous signal in MALDI, yet it also introduces a dilution effect of the analyte in the matrix crystal. This dilution effect, while effective in providing a signal that continues over many laser pulses, is ultimately detrimental to achieving the highest level of sensitivity.

The length of signal duration is also quite different between nanostructure-based laser desorption/ionization and MALDI. Typically NIMS generates a signal for a significantly shorter number of laser pulses (3–50) whereas MALDI can generate a signal for hundreds if not thousands of laser shots before signal depletion occurs. The shorter signal duration characteristic of the nanostructured surfaces is likely due to the very different nature of the matrix-free nanostructure versus matrix-induced events that can occur by using MALDI. Since NIMS [14] are surface-induced phenomena, the generation of a signal is largely a 2D surface phenomenon versus 3D matrix crystals that depends on the nanosecond duration of the thermal and surface-restructuring events. Having a signal from a larger packet of ions in fewer laser shots provides a higher signal/noise ratio (S/N) since it contains a fixed amount of noise. When data is averaged over a larger number of spectra, the S/N only increases in proportion to the square root of the number of shots taken and the relatively low surface concentration in NIMS is quickly depleted. Therefore, averaging spectra from multiple laser shots ultimately results in a lower S/N than getting a larger burst of ions detected. This is analogous to LC-MS where increasing chromatographic resolution with techniques like UPLC or smaller ID columns like nano and capillary LC boosts sensitivity. Finally, an additional difference observed is in the laser energy used in NIMS ($\sim$10 mJ/cm^2) which is significantly lower than that used for MALDI ($\sim$40 mJ/cm^2 or higher). These lower energies can presumably reduce extraneous signal that can occur as a result of fragmentation of analyte, thereby minimizing the accumulation of background noise and improving the S/N.

5   Conclusion

The high detection sensitivity that can be obtained by using NIMS is the result of efficient ion generation from these surfaces as well as extremely low background noise. As discussed, modifying the surface with fluorous compounds is very important to achieve yoctomole sensitivity. In addition, engineering the nanostructures could further enhance the detection sensitivity. For example, Vertes et al. demonstrated that nanofabrication of ordered monolithic silicon nanostructures such as NAPA, with optimized array geometries (including height and diameter of nanopost and post-to-post distance), has the potential to improve the detection sensitivity [14, 21]. Theorizing that the optimization of the array geometries enhances the nanostructure-laser interaction, therefore improving ion production; NAPA was capable of detecting $\sim$800 zmol of verapamil [21]. Therefore the combination of ordered nanostructured surfaces with fluorous surface modifications could further improve detection sensitivity beyond what has been observed thus far.

Currently, manual deposition is the most commonly used approach for sample deposition in nanostructure-based desorption/ionization MS experiments. In these cases deposition quantities typically range from 0.1 to 0.5 µl of sample solution. Alternatively, the use of more accurate sample deposition techniques (e.g., acoustic deposition) that effectively reduce the deposited sample volume could concentrate the sample to an even smaller spot size, and improve detection sensitivity. Acoustic deposition is capable of precisely depositing $\sim$100 pl sized droplets onto a surface with spot size as low as 60 µm [22]. Therefore, the combination of ultra-fine sample deposition techniques with the concentrating effect of a hydrophobic nanostructured surface may provide another possible way to further improve the sensitivity.

The biological implications of ultrahigh detection sensitivity are especially significant given its potential application to single-cell analysis [23, 24]. One significant application would be the ability to observe single-cell heterogeneity and elucidate the role that each cell plays in the function of a biological system. The size of a cell is typically 1–100 µm, with a volume of $\sim$30 fl. With the concentrations of major metabolites in cells in the attomole range [14], nanostructure-based desorption/ionization mass spectrometry exhibits a limit of detection down to yoctomole level, making metabolic imaging of single cells (i.e., intracellular metabolite biodistributions) possible to explore. Given the importance of ultrahigh detection sensitivity for single-cell analysis, nanostructure-based desorption/ionization mass spectrometry could ultimately play an important role in these analyses, providing new insights into cellular biology.

Acknowledgments

This work conducted by ENIGMA-Ecosystems and Networks Integrated with Genes and Molecular Assemblies was supported by the Office of Science, Office of Biological and Environmental Research, of the US Department of Energy under Contract No. DE-AC02-05CH11231. This work was also supported by the California Institute of Regenerative Medicine Grant TR1-01219 and the National Institutes of Health grants R24 EY017540-04, P30 MH062261-10, and P01 DA026146-02. Financial support was also received from the Department of Energy grants FG02-07ER64325 and DE-AC0205CH11231.

References

1. Thomson JJ (1910) Rays of positive electricity. Phil Mag 20:752–767

2. Karas M, Bachmann D, Bahr U, Hillenkamp F (1987) Matrix-assisted ultraviolet laser desorption of non-volatile compounds. Int J Mass Spectrom Ion Process 78:53–68

3. Karas M, Hillenkamp F (1988) Laser desorption ionization of proteins with molecular masses exceeding 10,000 daltons. Anal Chem 60:2299–2301

4. Wei J, Buriak JM, Siuzdak G (1999) Desorption-ionization mass spectrometry on porous silicon. Nature 399:243–246

5. Trauger SA, Go EP, Shen ZX, Apon JV, Compton BJ, Bouvier ESP, Finn MG, Siuzdak G (2004) High sensitivity and analyte capture with desorption/ionization mass spectrometry on silylated porous silicon. Anal Chem 76:4484–4489

6. Nordstrom A, Apon JV, Uritboonthal W, Go EP, Siuzdak G (2006) Surfactant-enhanced desorption/ionization on silicon mass spectrometry. Anal Chem 78:272–278

7. Northen TR, Yanes O, Northen MT, Marrinucci D, Uritboonthai W, Apon J, Golledge SL, Nordstrom A, Siuzdak G (2007) Clathrate nanostructures for mass spectrometry. Nature 449:1033–1036

8. Yanes O, Woo HK, Northen TR, Oppenheimer SR, Shriver L, Apon J, Estrada MN, Potchoiba MJ, Steenwyk R, Manchester M, Siuzdak G (2009) Nanostructure initiator mass spectrometry: tissue imaging and direct biofluid analysis. Anal Chem 81:2969–2975

9. Patti GJ, Woo HK, Yanes O, Shriver L, Thomas D, Uritboonthai W, Apon JV, Steenwyk R, Manchester M, Siuzdak G (2010) Detection of carbohydrates and steroids by cation-enhanced nanostructure-initiator mass spectrometry (NIMS) for biofluid analysis and tissue imaging. Anal Chem 82:121–128

10. Greving MP, Patti GJ, Siuzdak G (2011) Nanostructure-initiator mass spectrometry metabolite analysis and imaging. Anal Chem 83:2–7

11. Kruse RA, Li X, Bohn PW, Sweedler JV (2001) Experimental factors controlling analyte ion generation in laser desorption/ionization mass spectrometry on porous silicon. Anal Chem 73:3639–3645

12. Northen T, Woo H-K, Northen M, Nordström A, Uritboonthail W, Turner K, Siuzdak G

(2007) High surface area of porous silicon drives desorption of intact molecules. J Am Soc Mass Spectrom 18:1945–1949

13. Peterson DS (2007) Matrix-free methods for laser desorption/ionization mass spectrometry. Mass Spectrom Rev 26:19–34

14. Stolee JA, Walker BN, Zorba V, Russo RE, Vertes A (2012) Laser-nanostructure interactions for ion production. Phys Chem Chem Phys 14:8453–8471

15. Go EP, Apon JV, Luo GH, Saghatelian A, Daniels RH, Sahi V, Dubrow R, Cravatt BF, Vertes A, Siuzdak G (2005) Desorption/ionization on silicon nanowires. Anal Chem 77:1641–1646

16. Walker BN, Stolee JA, Pickel DL, Retterer ST, Vertes A (2010) Tailored silicon nanopost arrays for resonant nanophotonic ion production. J Phys Chem C 114:4835–4840

17. Chen Y, Vertes A (2006) Adjustable fragmentation in laser desorption/ionization from laser-induced silicon microcolumn arrays. Anal Chem 78:5835–5844

18. Two different individuals (S. Trauger and J. Apon) performed the experiments independently.

19. Two different individuals (W. Uritboonthai and O. Yanes) performed the experiments independently.

20. Keller BO, Li L (2001) Detection of 25,000 molecules of substance P by MALDI-TOF mass spectrometry and investigations into the fundamental limits of detection in MALDI. J Am Soc Mass Spectrom 12:1055–1063

21. Walker BN, Stolee JA, Vertes A (2012) Nanophotonic ionization for ultratrace and single-cell analysis by mass spectrometry. Anal Chem 84:7756–7762

22. Aerni H-R, Cornett DS, Caprioli RM (2005) Automated acoustic matrix deposition for MALDI sample preparation. Anal Chem 78:827–834

23. Trouillon R, Passarelli MK, Wang J, Kurczy ME, Ewing AG (2012) Chemical analysis of single cells. Anal Chem 85:522–542

24. O'Brien PJ, Lee M, Spilker ME, Zhang C, Yan Z, Nicholls TC, Li W, Johnson CH, Patti GJ, Siuzdak G (2013) Monitoring metabolic responses to chemotherapy in single cells and tumors using nanostructure-initiator mass spectrometry (NIMS) imaging. Cancer Metab 1:4

Chapter 15

Nanostructure-Initiator Mass Spectrometry (NIMS) for Molecular Mapping of Animal Tissues

Tara N. Moening, Victoria L. Brown, and Lin He

Abstract

Nanostructure-initiator mass spectrometry (NIMS) is an established method for sensitive detection of small molecules in complex samples. It is based on the optimal combination of a porous Si substrate and a carefully selected polymer coating to allow certain analytes of interest to be concentrated on the substrate for effective ionization with minimal background interference from conventional organic matrices. The previous chapter has detailed the history and current state of the art of the technique in small-molecule profiling and imaging applications. We describe here a simple step-by-step protocol for substrate fabrication and sample preparation that provides a starting point for the technique to be adapted and optimized for 2-D biological imaging applications.

Key words Nanostructure-initiator mass spectrometry (NIMS), Mass spectrometry imaging, Metabolites

1 Introduction

Matrix-assisted laser desorption/ionization (MALDI) has had improvements, adjustments, and parallel techniques added to its repertoire since its birth in the 1980s [1]. One such parallel technique is surface-assisted laser desorption/ionization (SALDI) which typically uses inorganic materials to help transfer energy provided by the laser to desorb and ionize the analyte [2, 3]. With SALDI, no matrix is used which is beneficial in analyzing small molecules such as metabolites. A recently introduced SALDI technique, nanostructure-initiator mass spectrometry (NIMS), has been proven to be quite successful by employing nanoporous etched silicon with a fluorinated polymer coating, the so-called initiator, to aid in desorption and ionization of the analyte of interest [3]. A brief overview of NIMS and its capability in small-molecule detection is discussed by its inventors, Siuzdak and his co-workers, in the previous chapter.

Lin He (ed.), *Mass Spectrometry Imaging of Small Molecules*, Methods in Molecular Biology, vol. 1203, DOI 10.1007/978-1-4939-1357-2_15, © Springer Science+Business Media New York 2015

In practice, successful conduct of NIMS experiments critically relies on the quality of the pSi substrate prepared and the proper initiator selected for the targeted applications. For example, depending on the nature of the analytes to be studies, a more hydrophobic, or more hydrophilic, coating is preferred to reduce the interaction between the surface coating and the analytes. The solubility of the initiator in a solvent that can be easily spread into pSi pores is another factor to be considered. Just as it is important to choose an appropriate matrix for conventional MALDI, the selection of an appropriate initiator is critical for successful NIMS measurements. Siuzdak and co-workers reported a list of possible initiators that could be used as a coating of the pSi surface [3–6]. Many of the initiators reported to date are hydrophobic in nature. With this hydrophobic surface, most water-based analyte solutions take a smaller contact angle with the surface and allow the analyte to be concentrated, which lead to greatly increased signal intensities, thus decreasing the limit of detection. NIMS has shown to achieve detection as low as the yoctomole range [3]. Additional factors to be optimized in NIMS measurements include solvent selection, initiator incubation time, and surface washes prior to analyte deposition.

In this chapter, we describe a generic experimental protocol using NIMS for mass spectrometric imaging of small molecules on a 2-dimensional surface, which provides a starting point for metabolite imaging of complex biological samples. Modifications may be needed in order to achieve optimal results.

2   Materials

2.1   Preparation of NIMS Substrate

1. N-type Sb-doped (100) single-crystalline silicon wafers at 0.005–0.02 Ω/cm (Silicon Sense, Inc.).

2. Piranha cleaning solution: 98 % sulfuric acid:30 % hydrogen peroxide (2:1 v/v) (*see* **Note 1**).

3. 25 % Hydrofluoric acid (HF) etching solution: 49 % HF:95 % ethanol (1:1 v/v).

4. Bis(heptadecafluoro-1,1,2,2-tetrahydrodecyl)tetramethyldisiloxane (BisF17).

5. Graduated cylinder, glass petri dish, HF-resistant tweezers, glass rod.

6. Teflon etching cell (*see* Fig. 1).

7. MALDI standard solution and matrix: Nominally 0.01 mM of angiotensin I, angiotensin III, and bradykinin, respectively, in water. 10 mg/mL of 2,5-dihydroxybenzoic acid (DHB) dissolved in 50:50:0.1 ACN:H_2O:TFA. The peptides and matrix were mixed at a 1:1 v/v ratio.

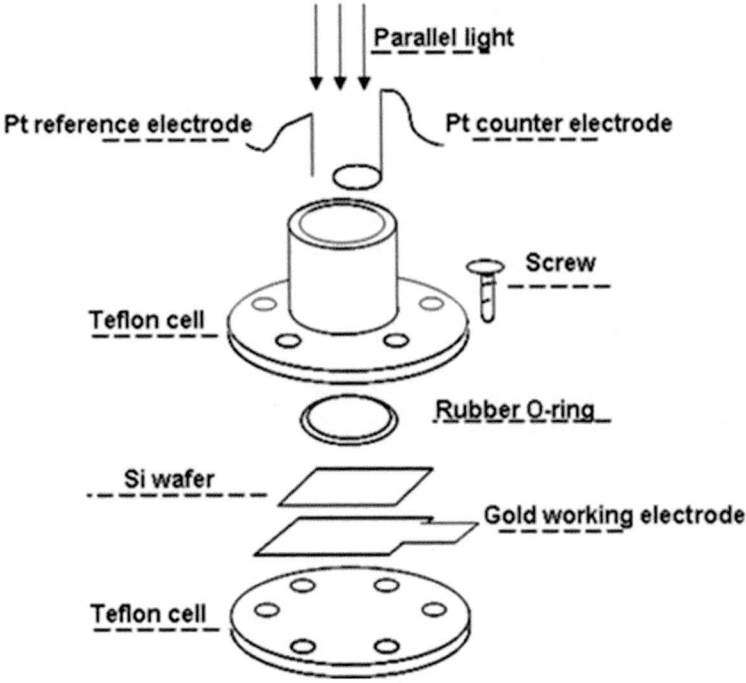

Fig. 1 Schematic drawing of an electrochemical etching cell for porous Si preparation. Adapted with permission from ref. 7

2.2 Tissue Section Preparation

1. Optical cutting temperature (OCT) compound.
2. Leica Surgipath DB80 HS Premium High Profile Disposable Microtome Blade (Leica Biosystems, Buffalo Grove, IL, USA).
3. Mouse brain tissue samples.
4. Double-sided conductive tape.

2.3 Instrumentation

1. EG&G Princeton Potentiostat, Model 273.
2. Leica CM1950 cryostat (Leica Biosystems, Buffalo Grove, IL, USA).
3. AB Sciex MALDI-TOF/TOF 5800 mass spectrometer (Framingham, MA).
4. TOF/TOF Series Explorer Software V4.1.0.
5. AB Sciex TOF/TOF Imaging Acquisition Software 1.0.

3 Methods

3.1 Preparation of NIMS Substrate

1. Cut silicon wafers into 1-cm^2 square-shaped chips and place in the piranha solution for 30 min to remove organic contaminants. Wash the Si chips with DI water and dry with N_2.

2. Assemble Si chip in an anodic etching cell made of Teflon, as shown in Fig. 1. A three-electrode system was used for surface etching where a Au working electrode was placed under the Si chip, and a Pt counter electrode and a Pt reference electrode were placed above the surface. Clamp the cell tight before filling with the HF etching solution (25 %). Program an EG&G Princeton Potentiostat, Model 273, to control etching current and time. The complete setup was placed in a chemical hood during substrate fabrication (*see* **Note 2**).

3. Electrochemically etch the Si chip in 25 % HF solution for 30 min at a current density of 32 mA/cm². The waste solution containing HF was carefully transferred to a pre-labeled plastic waste bottle in the hood dedicated to HF. The setup was washed with 95 % ethanol before disassembly. The produced NIMS substrate was washed again with 95 % ethanol and dried with N_2 (*see* **Note 3**).

4. Place the NIMS substrate in an oven at 100 °C for 5 min. Remove the substrate from oven and place it in a glass petri dish to cool down to room temperature.

5. Place approximately 33 μL of neat BisF17 on the surface of the NIMS substrate and incubated for 30 min. Blow off excess BisF17 with N_2 and then place the substrate in the oven 100 °C for 5–8 s. Repeat this blowing off of excess initiator process for a total of three times. After blowing off all excess BisF17, the substrate was washed with a generous amount of THF and dried with N_2 (*see* **Note 4**).

6. The substrate was stored in a sealed petri dish at room temperature until needed.

3.2 Tissue Section Preparation

1. Snap freeze mouse brain samples in liquid N_2 immediately and store at −80 °C prior to usage (*see* **Note 5**).

2. Mount frozen brain samples onto the microtome plate by spotting a small amount of OCT compound onto the plate which will hold the tissue in place. The temperature was slowly brought up to −20 °C. NIMS substrate was cooled to −20 °C. The frozen tissue samples were then sectioned using the Leica cryostat equipped with a Leica disposable microtome blade into 5-μm-thick slices. The 5-μm-thick tissue slices were transferred via tweezers onto a dry, chilled NIMS substrate before being brought to room temperature in order for the tissue to adsorb to the NIMS surface.

3. If NIMS substrate with tissue sections is not analyzed immediately, it should be stored at −80 °C until analyzed.

4. Proper cleaning of the cryostat and disposal of cryostat blades should occur after cutting of tissue samples is completed.

3.3 Mass Spectrometry Imaging

1. An AB Sciex TOF/TOF 5800 mass spectrometer was used for the following description. Modification in instrumental conditions may be needed to achieve optimal imaging results when different instruments are used.

2. The mass spectrometer was equipped with a 1,000-Hz Nd:YAG laser with a fixed diameter of 70 μm. The laser intensity was adjusted to optimize performance (*see* **Note 6**).

3. The NIMS substrate with a piece of tissue section attached was mounted on a stainless steel MALDI plate with double-sided conductive tape.

4. The MALDI instrument was calibrated by placing 1 μL of a standard mixture of angiotensin I (m/z 1,296.48), angiotensin III (m/z 931.09), and bradykinin (m/z 1,060.21) with DHB on the side of the NIMS substrate where the Si wafer was not etched.

5. The instrument was operated in a positive ion mode, accumulating 200 laser shots at each location to yield one accumulated spectrum for each imaging pixel. The translational stage was operated at 100 μm stepwise.

6. Data acquisition was controlled by the TOF/TOF Series Explorer Software V4.1.0. The MSI parameters were controlled by Sciex TOF/TOF Imaging Acquisition Software 1.0.

7. 2-D ion maps were reconstructed using MSiReader V0.03 (NC State University, W.M. Keck FT-ICR Mass Spectrometry Laboratory). This software can be downloaded for free at www.msireader.com (*see* **Note 7**).

4   Notes

1. CAUTION: Slowly add hydrogen peroxide to sulfuric acid; avoid splashing or overheating of the solution.

2. During the etching process, gas bubbles were generated mildly, and continuously released. The Si chip slowly turned to darker grey; this suggests the formation of a porous surface (Fig. 2). Formation of porous features on Si is critically related to the current density applied and the etching time.

3. The current density and etching time should be optimized depending on the electrochemical cell setup as well as the type of Si chips used (i.e., doping type and doping level). The optimal etching conditions should be optimized by the MS results.

4. When placing the porous Si in the oven, it is common to see excess BisF17 coming out of the pores, causing the surface to look wet. To avoid contamination of the oven, it is recommended

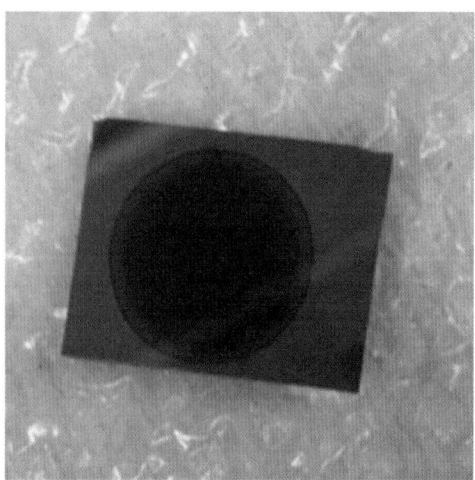

Fig. 2 A representative NIMS substrate under natural light. The *dark grey center area* is where porous Si features locate

to place the substrate in a glass petri dish or some glass support. Always dry the substrate immediately after the THF wash with N_2 to make sure that all residual THF is removed from the surface.

5. Freeze and thaw of tissue samples is a critical step in preparation of tissue sections. The most common problem observed in frozen tissue sections is the ice crystal damage, which causes leaky tissues and blotchy tissue surface. The size of ice crystals is usually determined by the speed of the freezing process through the whole tissue.

6. Irradiation energy is adjusted to achieve optimal MS performance prior to sample imaging. It is expected to be significantly lower than the energy needed for traditional MALDI, but it is common that the laser flux used in point checking is lower than what is needed in 2-D imaging experiment. A small-scale 2-D imaging testing is always recommended before any tissue imaging experiments to avoid waste of time and samples.

7. A representative NIMS imaging data is shown in Fig. 3 where a 5 μM dipalmitoylphosphatidylcholine (DPPC) in chloroform was spotted on a NIMS substrate. The shown image corresponds to the headgroup ions of DPPC ($m/z = 184.0$). *See* Chapter 14 for more complex tissue sample images.

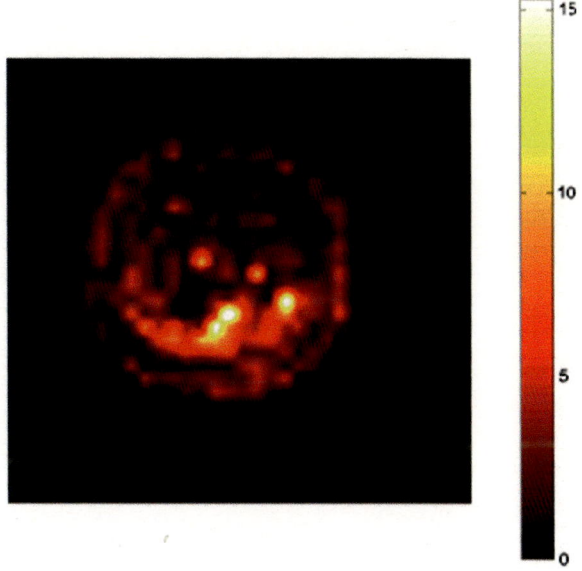

Fig. 3 A representative NIMS MS image collected with 5 μM dipalmitoylphosphatidylcholine (DPPC) spotted. The image shown is at $m/z = 184.0$ (i.e., DPPC headgroup). The scale bar on the *right* shows the corresponding ion intensity

References

1. Karas M, Bachmann D, Bahr U, Hillenkamp F (1987) Matrix-assisted ultraviolet laser desorption of non-volatile compounds. Int J Mass Spectrom Ion Process 78:53–68
2. Wei J, Buriak JM, Siuzdak G (1999) Desorption-ionization mass spectrometry on porous silicon. Nature 399:243–246
3. Woo H, Northen TR, Yanes O, Siuzdak G (2008) Nanostructure-initiator mass spectrometry: a protocol for preparing and applying NIMS surfaces for high-sensitivity mass analysis. Nature 3:1341–1349
4. Calavia R, Annanouch FE, Correig X, Yanes O (2012) Nanostructure initiator mass spectrometry for tissue imaging in metabolomics: future prospects and perspectives. J Proteom 75:5061–5068
5. Greving MP, Patti GJ, Siuzdak G (2011) Nanostructure-initiator mass spectrometry metabolite analysis and imaging. Anal Chem 83:2–7
6. Yanes O, Woo H, Northen TR, Oppenheimer SR, Shriver L, Apon J, Estrada MN, Potchoiba MJ, Steenwyk R, Manchester M, Siuzdak G (2009) Nanostructure initiator mass spectrometry: tissue imaging and direct biofluid analysis. Anal Chem 81:2969–2975
7. Liu Q et al (2010) Mass spectrometry imaging. In: Rubakhin SS, Sweedler JV (eds) Methods in molecular biology, vol 656. Humana Press, Totowa, NJ, p 243

Chapter 16

Nanoparticle-Assisted Laser Desorption/Ionization for Metabolite Imaging

Michihiko Waki, Eiji Sugiyama, Takeshi Kondo, Keigo Sano, and Mitsutoshi Setou

Abstract

Matrix-assisted laser desorption/ionization imaging mass spectrometry (MALDI-IMS) has enabled the spatial analysis of various molecules, including peptides, nucleic acids, lipids, and drug molecules. To expand the capabilities of MALDI-IMS, we have established an imaging technique using metal nanoparticles (NPs) to visualize metabolites, termed nanoparticle-assisted laser desorption/ionization imaging mass spectrometry (nano-PALDI-IMS). By utilizing Ag-, Fe-, Au-, and TiO_2-derived NPs, we have succeeded in visualizing various metabolites, including fatty acid and glycosphingolipids, with higher sensitivity and spatial resolution than conventional techniques. Herein, we describe the practical experimental procedures and methods associated with nano-PALDI-IMS for the visualization of these molecules.

Key words Imaging mass spectrometry, Nanoparticles, Matrix, Fatty acids, Glycosphingolipids, Metabolites

1 Introduction

Matrix-assisted laser desorption/ionization imaging mass spectrometry (MALDI-IMS) using organic matrixes is a popular analytical technique and its applications are increasing in the fields of biology and medicine [1–3]. The advantages of MALDI-IMS over conventional molecular imaging techniques are as follows [4, 5]. (1) No labeling of the molecules to be analyzed is required. (2) Multiple molecules within the measuring mass range are simultaneously visualized. (3) Molecular structure can be analyzed by multistage MS (MS^n) performed directly on the tissue section, thus enabling analysis of unknown molecules. These advantages differentiate MALDI-IMS from conventional imaging methods, such as immunohistochemistry and in situ hybridization, and make it particularly well suited for use in metabolome studies (Fig. 1).

Lin He (ed.), *Mass Spectrometry Imaging of Small Molecules*, Methods in Molecular Biology, vol. 1203, DOI 10.1007/978-1-4939-1357-2_16, © Springer Science+Business Media New York 2015

molecular diversity

	Representative molecular imaging method in tissue or cells	probes	selectivity	allow simultaneous imaging
DNA	FISH (fluorescence *in situ* hybridization)	oligo nucleotide probe	targeted	+
RNA	*in situ* hybridization	oligo nucleotide probe	targeted	+
protein	immunohistochemistry, green fluorescent protein-fused protein	antibody	targeted	+
metabolite (especially lipids)	imaging mass spectrometry	-	targeted/non-targeted	+++

Fig. 1 Representative molecular imaging methods. Reprinted from [31] with permission from Springer Science

Efforts to improve the ionization process in conventional MALDI-IMS are ongoing. Strategies to improve this process involve the following: (1) expand detectable molecular species, which is constrained by the physicochemical character of the organic matrix; (2) reduce the effect of low-molecular-weight fragments derived from the matrix material in the case of organic matrixes, which interfere with the spectra of low-molecular-weight (less than approximately 500 Da) analyte molecules in MALDI; and (3) improve spatial resolution in MALDI, which is restricted by the size of the crystals in the organic matrix [6].

A number of techniques have been investigated to address these issues [7–9]. For example, matrix-free methods, such as desorption/ionization on porous silicon (DIOS), which employs an etched silicon wafer, are capable of analyzing low-molecular-weight molecules [10]. Porous, monolithic materials have also been found to efficiently ionize small molecules [11]. Such methods typically utilize the materials' ability to absorb laser energy and transfer it to the analyte material [12]. Using an organic matrix, a sol–gel polymeric structure into which dihydroxybenzoic acid (DHB) was covalently incorporated was reported to be used in MS analysis without the background of matrix interference spectra [13].

In this study, we developed nanoparticle laser desorption/ionization imaging mass spectrometry (nano-PALDI-IMS), a robust technique that is inexpensive, quick, and easy to perform, and that does not require specialty substrates [14, 15]. Nano-PALDI-IMS utilizes nanoparticles (NPs) composed of metal and nonmetal materials [16–18] or metals bound to fatty acid chains [14, 15], and these NPs are sprayed on tissue sections instead of organic matrixes, as in conventional MALDI-IMS (Fig. 2). We have succeeded in visualization of a vast number of metabolite species, including lipids, by employing this methodology. In this chapter, we present practical experimental nano-PALDI-IMS procedures that use four types of NPs—Ag-, Fe-, Au-, and TiO_2-derived NPs—for metabolite analysis. Protocols for the preparation of NPs and their application method for IMS analysis are presented. As analysis examples, we also present our recent results of nano-PALDI-IMS analysis of metabolites in tissue specimens.

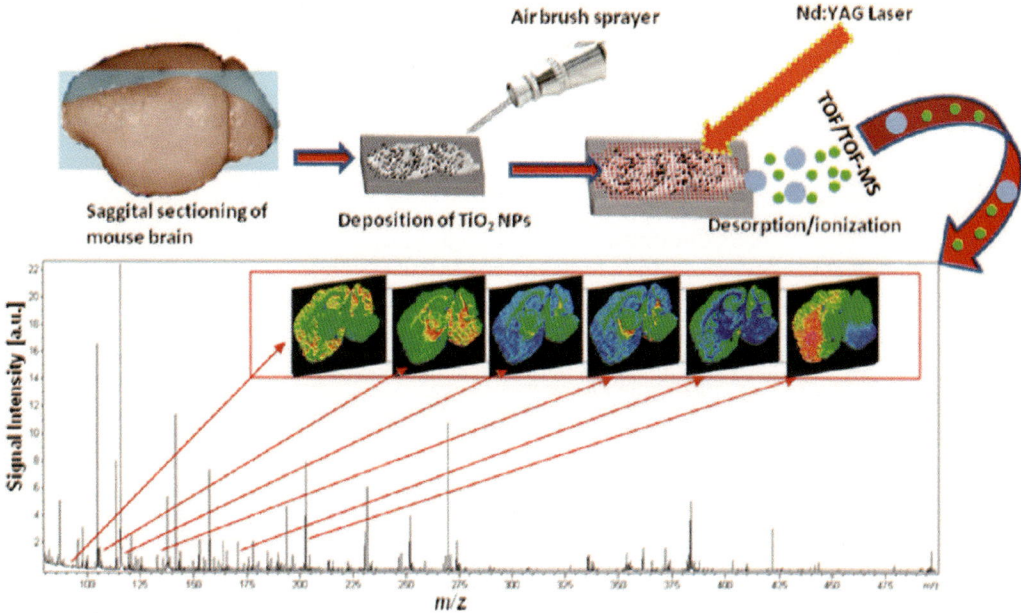

Fig. 2 Overall workflow for nano-PALDI-IMS. Tissues were sliced with a cryostat, thaw-mounted on ITO-coated glass slides, NPs were applied via an airbrush sprayer, and then, the tissue surface was analyzed by nano-PALDI-IMS. Reprinted from [16] with permission from American Chemical Society

2 Materials

2.1 Animals

1. C57BL/6 J mice were obtained directly from a commercial breeder.

2.2 Chemicals

2. Indium-tin-oxide (ITO)-coated glass slides.

3. Optimal-cutting-temperature (OCT) polymer.

4. Dry ice.

5. Liquid nitrogen.

   The following materials were used to prepare Ag NPs:

6. *n*-Tetradecanoic acid sodium salt (n-$C_{13}H_{27}COONa$): 0.15 M solution in 50 % ethanol, 1 L.

7. Stearylamine (octadecylamine).

8. NaOH: 1 M solution in water, 150 mL.

9. $AgNO_3$: 1 M solution in water, 165 mL.

   The following materials were used to prepare Fe NPs:

10. $FeCl_2 \cdot 4H_2O$ (iron(II) chloride tetrahydride): 100 mM solution in water, 20 mL.

11. 3-Aminopropyltriethoxysilane (APTES, γ-aminopropyltriethoxysilane).

12. Sodium acetate: 40 mM solution in methanol, 1 mL.

The following materials were used to prepare Au NPs:

13. HAuCl$_4$·4H$_2$O (hydrogen tetrachloroaurate(III) tetrahydride): 1 M solution in methanol, 5 mL.

14. Dimethyl sulfide.

15. Oleylamine (octadecylamine).

The following materials were used to prepare TiO$_2$ NPs:

16. Titanium(IV) *n*-butoxide (titanium(IV)tetra(1-butanolate)).

17. Nitric acid.

18. Diammonium hydrogen citrate.

19. Citric acid.

2.3 Instruments

20. Cryomicrotome; Leica CM1950 cryostat (Leica Microsystems, Nussloch, Germany).

21. Ultraflex II (Bruker Daltonics, Karlsruhe, Germany).

22. Qstar Elite (AB SCIEX, Framingham, MA, USA).

23. Transmission electron microscope (TEM) JEM-1010 and JEM-1230 (JEOL Ltd., Tokyo, Japan).

24. −80 °C freezer.

25. Magnetic stirrer with heating.

26. Water bath.

27. Desiccator containing a silica gel canister.

28. Porcelain mortar.

29. Airbrush with a 0.2 mm nozzle caliber (Procon Boy FWA Platinum; GSI Creos, Tokyo, Japan).

3 Methods

Metal NPs have enabled the MS analysis of a diverse array of molecular species [19]. Our experimental results suggest that target molecules suitable for each type of NPs in nano-PALDI-IMS are as follows: Ag NPs, Fe NPs, Au NPs, and TiO$_2$ NPs are, respectively, suitable for the analysis of fatty acids in negative ion mode, negatively charged lipids such as sulfatides and phosphatidylserines in negative ion mode, sulfatides and gangliosides in negative ion mode, and low-molecular-weight metabolites (LMWMs) (80–500 *m/z*) in positive ion mode.

3.1 NP Preparation

3.1.1 Ag NPs

The Ag NPs were prepared by a coupling reaction of *n*-C$_{13}$H$_{27}$COOAg with stearylamine.

1. The NaOH solution was added to the *n*-C$_{13}$H$_{27}$COOH solution.

2. The $AgNO_3$ solution was added, and the white precipitate of n-$C_{13}H_{27}COOAg$ was collected and dried under reduced pressure at 60 °C.

3. 1 mmol of n-$C_{13}H_{27}COOAg$ and 1 mmol of stearylamine were mixed in a one-necked flask and heated at 120 °C for 5 h.

4. After cooling to 80 °C, methanol was added and the Ag NP precipitate was collected by filtration. The Ag NPs were washed with methanol and dried under vacuum.

5. A 50 mg/mL solution was prepared in a microtube by adding 500 μL of hexane to the Ag NPs (*see* **Note 1**).

3.1.2 Fe NPs

Magnetic Fe NPs were prepared by the coupling reaction between $FeCl_2$ with the ethoxy-group-containing APTES.

1. The $FeCl_2 \cdot 4H_2O$ solution was mixed with 20 mL of APTES and stirred for 1 h at room temperature.

2. The Fe NP precipitate was washed several times with water, dried at 80 °C, and ground in a porcelain mortar.

3. 10 mg of Fe NPs were dispersed in a sodium acetate solution in a microtube.

3.1.3 Au NPs

Au NPs were prepared by the thermolysis reaction of gold(I) sulfide with $AuCl(SMe_2)$ and the subsequent substitution of SMe_2 with oleylamine.

1. $AuCl(SMe_2)$, a white precipitate, was prepared by dissolving dimethylsulfide (10 mmol) in a $HAuCl_4 \cdot 4H_2O$ solution and heating [20].

2. $AuCl(SMe_2)$ (295 mg, 1 mmol) and oleylamine (10 mmol) were mixed in a 10 mL flask by a magnetic stirrer.

3. The transparent mixture was gradually heated to 120 °C, at which time a homogeneous purple solution formed. The solution was kept at 120 °C for 1 h, and then cooled to room temperature.

4. 5 mL of acetone and 1 mL of methanol were added and mixed, and the resulting solution was centrifuged at $400 \times g$ for 5 min.

5. The Au NP precipitate was collected and dried under vacuum. Then, a 1 mL Au NP solution at 50 mg/mL was prepared in hexane in a microtube.

3.1.4 TiO₂ NPs

TiO_2 NPs were synthesized by the acid-catalyzed hydrolysis of titanium(IV) butoxide, followed by condensation [21].

1. 17 mL of titanium(IV) n-butoxide and 8 mL of ethanol were mixed by stirring for 10 min at room temperature.

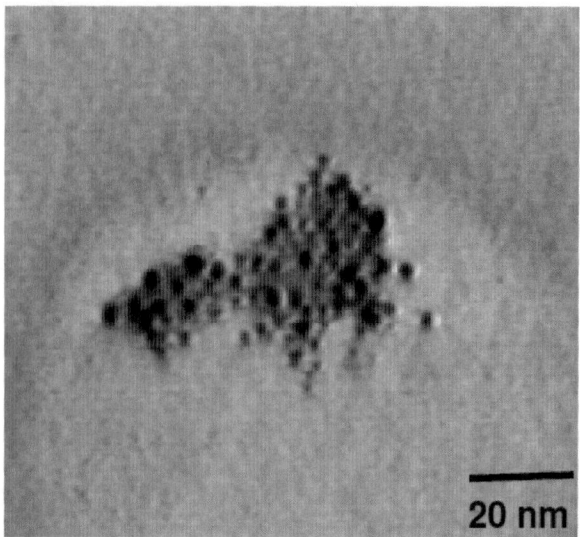

Fig. 3 Transmission electron microscopy image of TiO$_2$ NPs. Reprinted from [16] with permission from American Chemical Society

2. 375 µL of nitric acid was added dropwise to the titanium (IV) *n*-butoxide solution, which was cooled in an ice-containing water bath under stirring. TiO$_2$ NPs precipitated from this solution.

3. A 9.7 mM TiO$_2$ NP solution was prepared by dissolving NPs into 1 mL of methanol containing 50 mM diammonium hydrogen citrate and 100 mM citric acid in a microtube (*see* **Note 2**).

3.1.5 Determination of Diameters of NPs

To deposit a thin layer of NPs on the target surface, NPs with sizes of less than 10 nm are required [16]. The NP mean diameters were determined by transmission electron microscopy (TEM; Fig. 3).

3.2 Animal Tissue Extraction

All experiments involving mice were conducted in accordance with the protocols approved by the animal care and use committee at the research institute.

1. Eight-week-old C57BL/6 J mice were humanely killed.

2. Their organs were surgically collected and rapidly frozen in liquid nitrogen or powdered dry ice (*see* **Note 3**). The liver and retina were used for Ag NP experiments. The brain was used for Au and TiO$_2$ NP experiments. The brains were collected within 1 min after sacrifice.

3.3 Preparation of Tissue Section

1. Tissues were affixed to an OCT polymer, taking care that no polymer was mixed into the tissue slices (Fig. 4 and *see* **Note 4**).

2. Frozen tissue sections with a thickness of 5–15 µm were prepared at –16 to –20 °C using a cryostat (*see* **Note 5**).

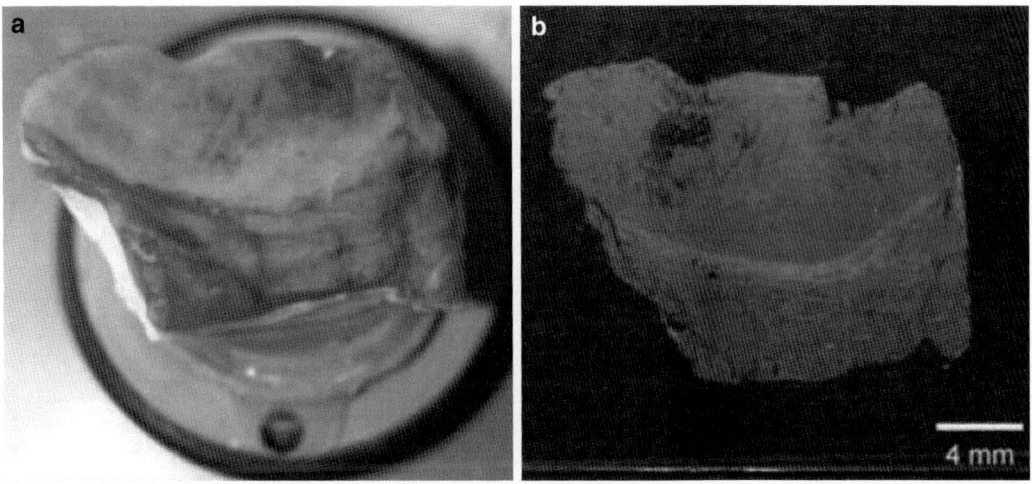

Fig. 4 Tissue sample on a cryostat sample holder. (**a**) The sample was affixed to the cryostat steel plate using a small amount of OCT. (**b**) Tissue section of 10 μm mounted onto a conductive glass slide for MALDI-IMS. Reprinted from [32] with permission from Springer Science

3. The sections were thaw-mounted onto ITO-coated glass slides.

4. The prepared sections were subjected to matrix application within 5 min (*see* **Note 4**). When the slides were not used immediately, they were put in a slide mailer, packed into Ziploc with silica gel, and stored at –80 °C. Frozen slides were dried in a desiccator prior to use.

3.4 Spraying of NPs

Prepared NP solution was sprayed onto the tissue surfaces with a 0.2-mm nozzle caliber airbrush (*see* **Note 6**, Fig. 5). The distance between the nozzle and glass slide was kept at approximately 15 cm.

3.5 Instrument Parameter Settings

Metabolite distribution was measured using TOF/TOF-IMS instruments equipped with Nd:YAG lasers; an Ultraflex II system was used for the Ag, Fe, and TiO$_2$ NP samples, and a QSTAR Elite system was used for the Au NP samples.

1. Before performing the analysis, the instrument was calibrated with DHB, bradykinin, and angiotensin II, at m/z 155.03 [M+H]$^+$, 757.40 [M+H]$^+$, and 1046.54 [M+H]$^+$ in positive ion mode, and 153.03 [M-H]$^-$, 755.40 [M-H]$^-$, and 1044.54 [M-H]$^-$ in negative ion mode.

2. The operating conditions, including the laser energy and detector gain, were optimized to maximize the signal intensity. Scan pitches were set at 10–70 μm in accord with the purpose of the experiment (*see* **Note 7**).

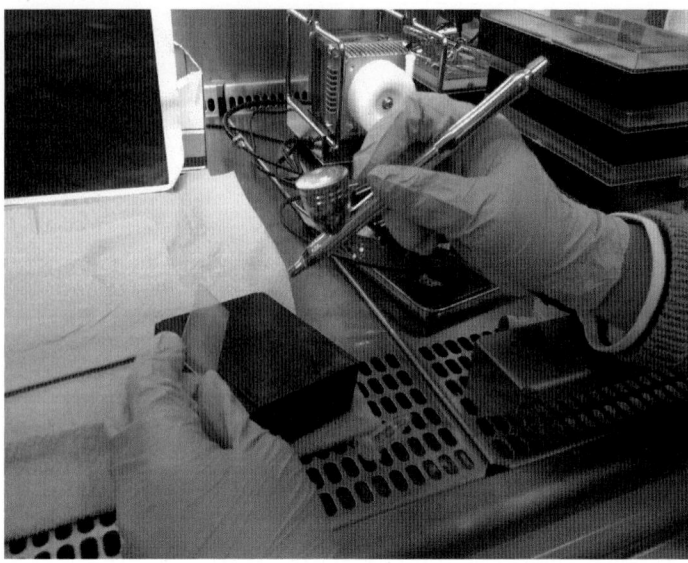

Fig. 5 NP solutions are sprayed onto the tissue samples using a 0.2 mm nozzle caliber airbrush

(a) Ag NPs: Negative ion and reflector modes; repetition rate of 200 Hz; laser irradiation of 200 shots per measurement point; mass range of 150–350 m/z.

(b) Fe NPs: Negative ion and reflector modes; laser irradiation of 100 shots in each spot; mass range of 450–1,500 m/z.

(c) Au NPs: Negative ion mode; repetition rate of 1,000 Hz; laser power of 60 %; mass range of 500–2,000 m/z.

(d) TiO$_2$ NPs: Positive ion mode; laser energy of 100 μJ; mass range of 80–500 m/z.

3. To confirm that the detected ion molecules were from the tissue analyte, an area that did not contain the tissue section was also irradiated using the laser (*see* **Note 8**).

3.6 Data Analysis

The ion images at arbitrary m/z values and mass spectra from regions of interest (ROI) were constructed using flexImaging and Biomap (Novartis, Basel, Switzerland) software programs. Mean signal intensities were obtained using tools in the software programs, and were utilized for quantitative analysis (*see* **Note 9**).

3.7 Identification of Ions

Identification of the detected MS peaks is often required if information on molecules that are detected in the analyzed organ or tissue is not well established.

1. On-tissue MS/MS analysis was performed to identify lipid structures, and capillary electrophoresis (CE)-MS was performed for the identification of LMWMs.

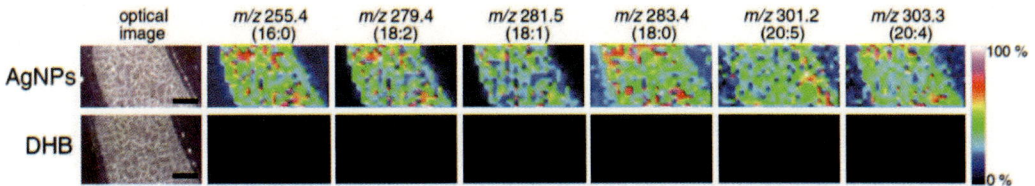

Fig. 6 Detection of fatty acids from mouse liver sections by nano-PALDI-IMS using Ag NPs and MALDI-IMS using DHB. The scale bars are 500 μm. Reprinted from [15] with permission from American Society for Mass Spectrometry

2. Peaks were assigned using the Human Metabolome Database (HMDB; Alberta, Canada; http://www.hmdb.ca/) and the Kyoto Encyclopedia of Genes and Genomes (KEGG; Kyoto, Japan; http://www.genome.jp/kegg/).

3. The assignments were validated by MS/MS and CE-MS measurements of standard compounds.

3.8 Ag NP-Based Nano-PALDI-IMS Analysis

In mammals, diverse fatty acids are synthesized, and these influence the development and pathological and physiological conditions of these species [22]. Specific types of fatty acids such as arachidonic acid (20:4) (AA), eicosapentaenoic acid (20:5) (EPA), and docosahexaenoic acid (22:6) (DHA) have been the focus of many investigations due to their immuno/thrombomodulatory activity and their effect on plasma membrane fluidity [23, 24].

Using nano-PALDI-IMS with Ag NPs, seven fatty acid species in mouse liver sections were detected with higher sensitivity as compared to MALDI-IMS with DHB (Fig. 6). Ion images were constructed with a spatial resolution of 10 μm (pitch) on a mouse retina section (Fig. 7). The ion image of EPA at 301.3 m/z was observed with biased intensity in some of the layers. AA and DHA, with m/z of 303.4 and 327.3, respectively, were detected predominantly in the retinal pigment epithelium. This observation is consistent with that presented in a previous report on the localization of phosphatidylcholine (16:0/22:6) (PC) and PC (18:0/22:6), which are major types of phospholipids in the plasma membrane [25, 26].

3.9 Fe NP- and Au NP-Based Nano-PALDI-IMS Analyses

Glycosphingolipids are amphiphilic molecules composed of hydrophilic carbohydrates and hydrophobic ceramides, and they are involved in biological processes such as cell proliferation and signaling [27, 28]. Among the glycosphingolipids, sulfatides are a class of sulfated galactosylceramides [29]. Using nano-PALDI-IMS with Fe NPs, sulfatides were successfully visualized in the dentate gyrus of mouse brain by discriminating layer structures [17, 18]. This approach enabled the analysis of sulfatides with high resolution at the cellular level (15 μm). Furthermore, the introduction of Au NPs enabled the visualization of sulfatides with enhanced signal

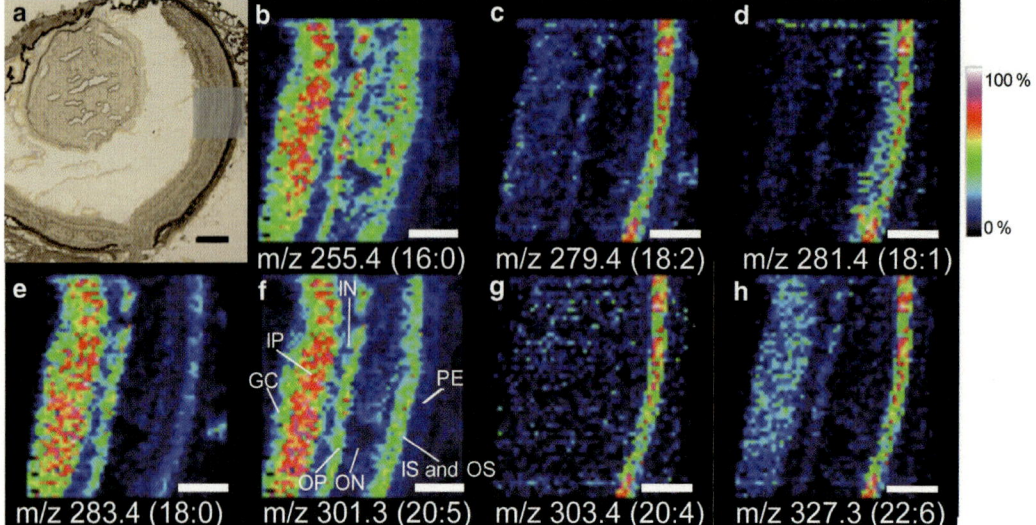

Fig. 7 Visualization of fatty acids in mouse retinal sections by nano-PALDI-IMS using Ag NPs. (**a**) The measurement area for the mouse retinal section is shown in *gray*. Reconstructed ion images of (**b**) palmitic acid (16:0), (**c**) linoleic acid (18:2), (**d**) oleic acid (18:1), (**e**) stearic acid (18:0), (**f**) EPA (20:5), (**g**) AA (20:4), and (**h**) DHA (22:6). The white scale bars are 100 μm. Reprinted from [15] with permission from American Society for Mass Spectrometry

intensity [14]. In total, ten major sulfatide species were visualized: d18:1/C18:0, d18:1/hC18:0, d18:1/C20:0, d18:1/hC20:0, d18:1/C22:0, d18:1/C24:1, d18:1/hC24:1, d18:1/hC24:0, d18:1/hC26:0, and d18:1/hC26:1 (Fig. 8). The average signal intensity of the molecules imaged with Au NPs was approximately 20 times higher than that of molecules imaged with DHB.

3.10 TiO₂ NP-Based Nano-PALDI-IMS Analysis

Endogenous LMWMs in brain have important roles as the products in molecular metabolism, source of energy production, or signal transducers [30]. By introducing TiO_2 NPs into mouse brain tissue, high-quality mass spectra free from matrix background signals were obtained, whereas spectra in the case of MALDI with DHB exhibited numerous peaks from DHB-derived adducts (Fig. 9). The distribution of molecules identified by MS/MS and/ or CE-MS analysis, using standard compounds, was reconstructed into two-dimensional pseudo-colored images (Fig. 10).

4   Notes

1. Ionization of molecules depends on the efficiency of complex formation among the analytes and NPs, which differs according to the analyte molecular species. It is therefore desirable that the effect of NP concentration on the intensity and number of detected target molecules is determined prior to detailed measurement.

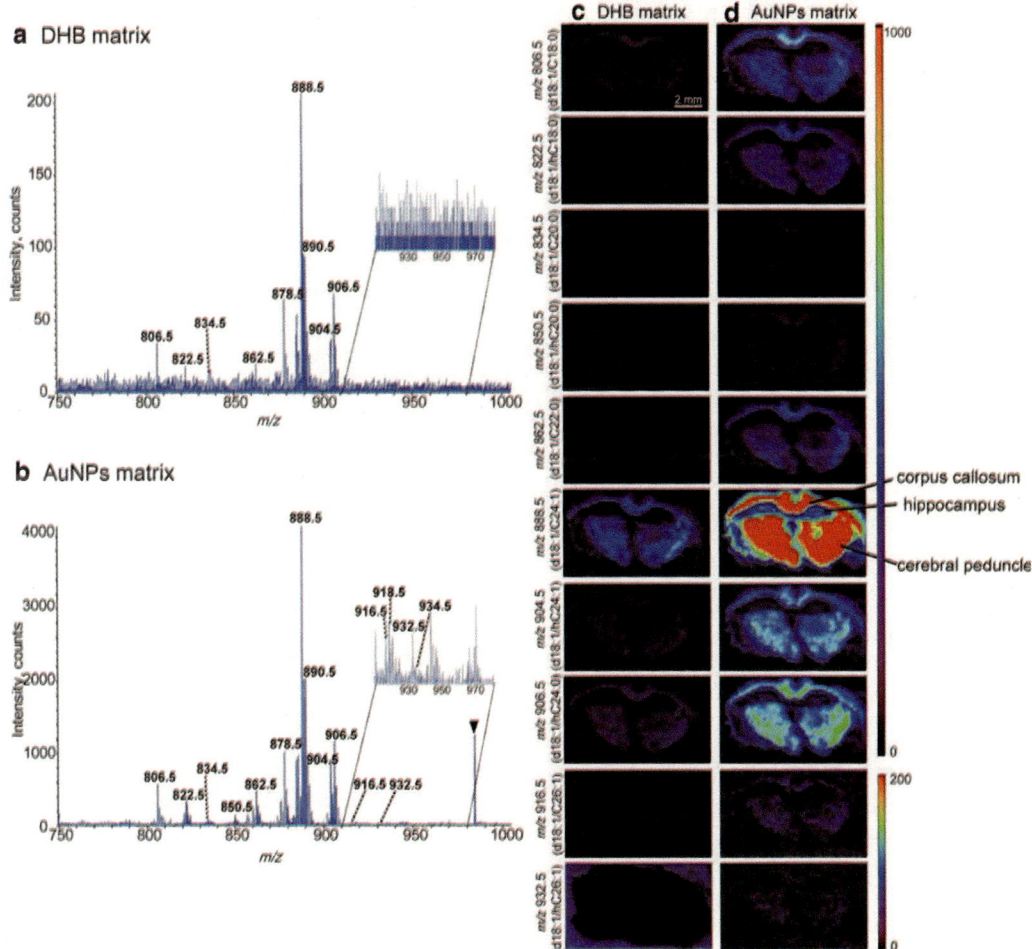

Fig. 8 Mass spectrum and ion images of sulfatide species by nano-PALDI-IMS using Au NPs. Mass spectra obtained using (**a**) DHB and (**b**) Au NPs, and the ion images of ten sulfatide species using (**c**) DHB and (**d**) Au NPs. Reprinted from [14] with permission from American Society for Mass Spectrometry

2. Concentrated hexane and methanol solutions were used for the extraction of lipids. Additional reagents can be involved in the solution to restrict ion additives [30]. In the study using TiO_2 NPs, citrate was added to promote protonation and predominant detection of $[M+H]^+$ [16].

3. Tissue can be rapidly frozen in liquid nitrogen. To maintain the fine structure of the organs without any damage, it is recommended that gradual freezing with powdered dry ice be performed. After freezing, the tissues should be stored at −80 °C until usage.

4. Polymer materials, such as OCT, are not recommended for embedding tissue samples because they suppress analyte ionization (Fig. 11). Samples of sufficient size (larger than the length of approximately 1 cm in any axis) do not require embedding. A small volume (which covers one edge of a sample)

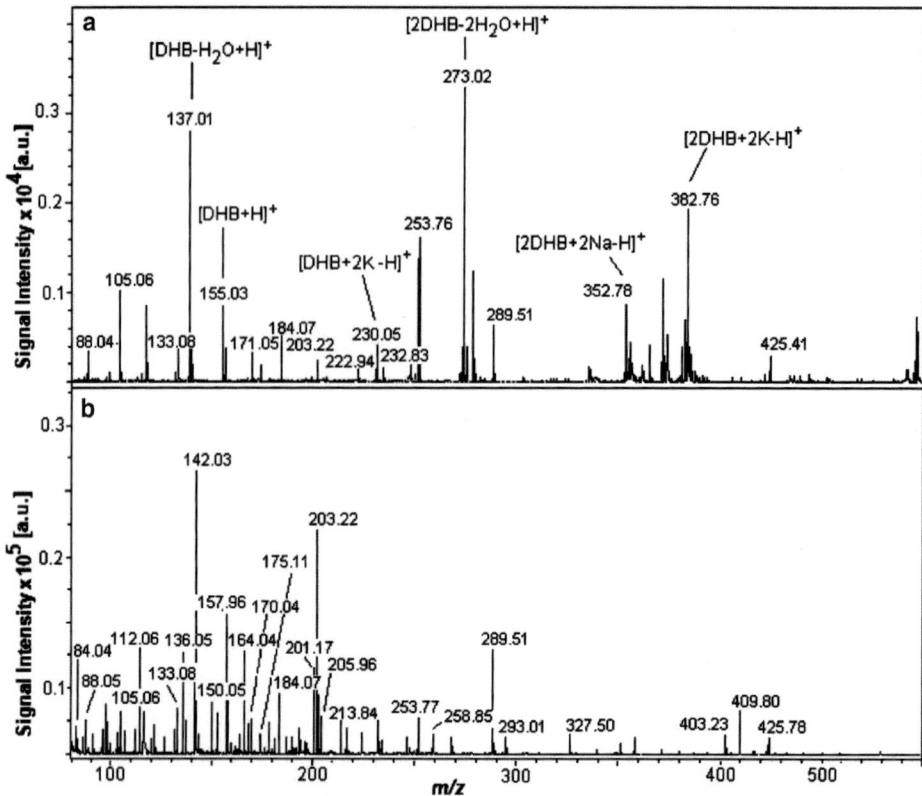

Fig. 9 Mass spectra of LMWM species from mouse brain tissue section obtained using (**a**) DHB in MALDI-IMS, and (**b**) TiO$_2$ NPs in nano-PALDI-IMS (**b**). Reprinted from [16] with permission from American Chemical Society

of OCT (this is just to settle the sample on the folder) should be placed on the folder, analyte specimen should be placed on it, and then, they should be placed in the cryostat until the sample gets fixed on the folder (Fig. 4). If the sample is too small to be set directly on the folder, it can be embedded in 2 % carboxymethyl cellulose (CMC). However, take care that frozen CMC is a hard material that generally makes sectioning a slightly difficult task.

5. The suitable temperature at which tissues are to be sectioned depends on the organ and analyte molecules. The results of previous immunohistochemical studies indicate that frozen organs can be analyzed. When lipids are analyzed, a temperature from –5 to –30 °C is preferred.

6. While spraying NPs, it is important to maintain a constant and weak airflow to avoid disrupting the tissue structure and heterogeneous distribution of NPs. In this study, the air spray with limiter was used to solve this problem. Prior to application to the sample, the appropriate amount should be sprayed on colored paper or plastic material and the drying time should be confirmed to be 1–2 s. It is desirable that the spraying is

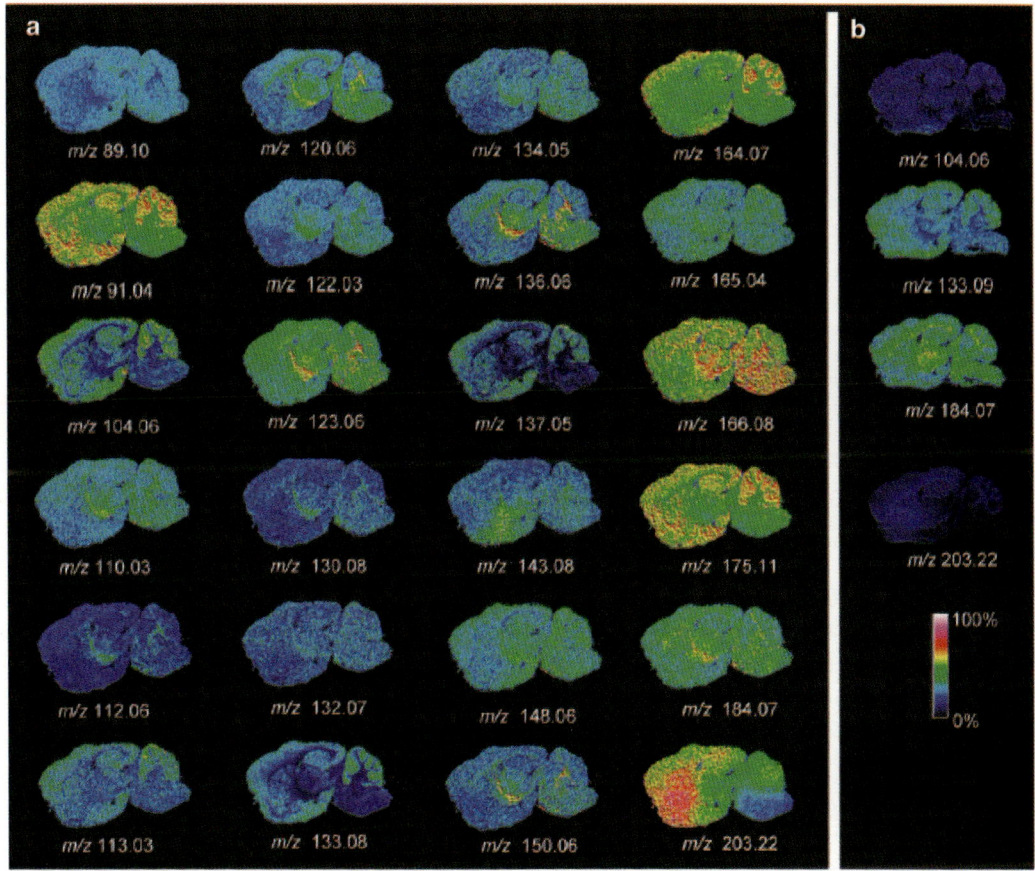

Fig. 10 Ion images of LMWM species in mouse brain tissue section obtained with (**a**) TiO$_2$ NPs in nano-PALDI-IMS, and (**b**) DHB in MALDI-IMS. Reprinted from [16] with permission from American Chemical Society

performed at constant room temperature and humidity to ensure experimental reproducibility. Also, it is important to consistently spray perpendicularly to the glass slide, to avoid biasing the sprayed volume between the distal and proximal point from an airbrush. In order to evaluate the distribution of NPs, an internal standard solution is applied onto the glass slides at multiple spots, and the variance of the standard's signal intensity is determined.

7. Because this method is not restricted by size of the matrix crystals, the irradiation pitch can be set to the instrument's minimum value. Users are expected to determine the optimal pitch, taking into account the measurement time and data storage requirements.

8. It is important to measure the area of glass slides coated only with NPs but not with tissue section as a negative control, because it is necessary to confirm that peaks that correspond to NPs in the course of laser irradiation and that are not related to the measured tissue section do not appear in mass spectra.

172   Michihiko Waki et al.

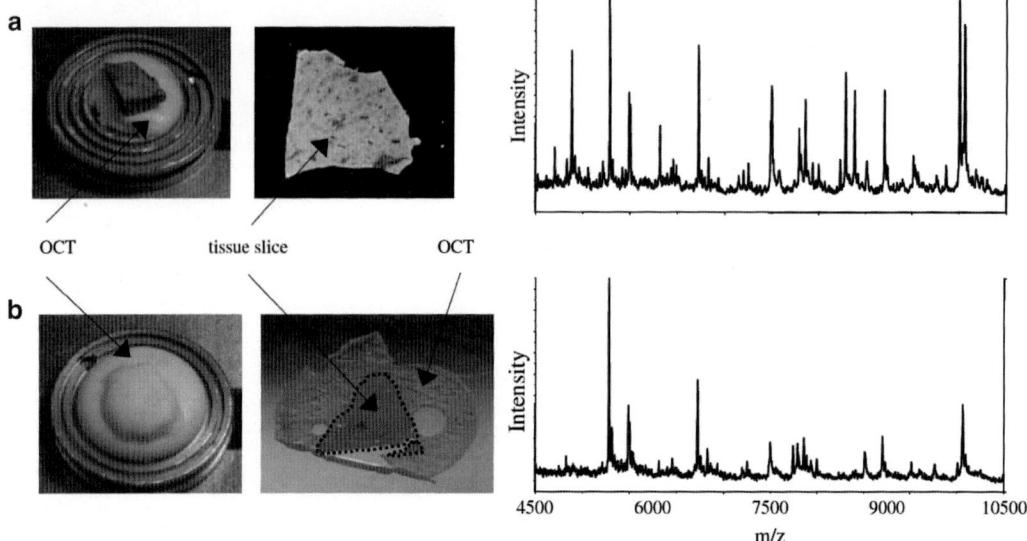

Fig. 11 Effect of OCT on signal intensity from rat liver tissue. (**a**) The optimal procedure involves using OCT to fix the tissue to the holder while avoiding contact of OCT with the sliced tissue; the resultant spectrum shows a number of intense signals. (**b**) The tissue was embedded in OCT and it is surrounded by OCT; the resultant spectrum contains approximately half of the signals observed in (**a**). Reprinted from [33] with permission from John Wiley & Sons, Ltd

9. Normalization of data is important in IMS to reduce variance of signal intensities among the spots on the section [30]. Total-ion-current (TIC)-based normalization was used in the case of Ag NP and Au NP experiments. This was performed by using the function in the flexImaging software (by selecting "normalize spectra" in the main window).

References

1. Schwamborn K, Caprioli RM (2010) Molecular imaging by mass spectrometry–looking beyond classical histology. Nat Rev Cancer 10(9): 639–646

2. Stoeckli M, Chaurand P, Hallahan DE et al (2001) Imaging mass spectrometry: a new technology for the analysis of protein expression in mammalian tissues. Nat Med 7(4): 493–496

3. Saito Y, Waki M, Hameed S et al (2012) Development of imaging mass spectrometry. Biol Pharm Bull 35(9):1417–1424

4. Shimma S, Sugiura Y, Hayasaka T et al (2008) Mass imaging and identification of biomolecules with MALDI-QIT-TOF-based system. Anal Chem 80(3):878–885

5. Cornett DS, Reyzer ML, Chaurand P et al (2007) MALDI imaging mass spectrometry:

molecular snapshots of biochemical systems. Nat Methods 4(10):828–833

6. Garden RW, Sweedler JV (2000) Heterogeneity within MALDI samples as revealed by mass spectrometric imaging. Anal Chem 72(1):30–36

7. Northen TR, Yanes O, Northen MT et al (2007) Clathrate nanostructures for mass spectrometry. Nature 449(7165):1033–1036

8. Cha S, Yeung ES (2007) Colloidal graphite-assisted laser desorption/ionization mass spectrometry and MSn of small molecules. 1. Imaging of cerebrosides directly from rat brain tissue. Anal Chem 79(6):2373–2385

9. Vidova V, Novak P, Strohalm M et al (2010) Laser desorption-ionization of lipid transfers: tissue mass spectrometry imaging without MALDI matrix. Anal Chem 82(12): 4994–4997

10. Wei J, Buriak JM, Siuzdak G (1999) Desorption-ionization mass spectrometry on porous silicon. Nature 399(6733):243–246

11. Peterson DS, Luo Q, Hilder EF et al (2004) Porous polymer monolith for surface-enhanced laser desorption/ionization time-of-flight mass spectrometry of small molecules. Rapid Commun Mass Spectrom 18(13):1504–1512

12. Tanaka K, Waki H, Ido Y et al (1988) Protein and polymer analyses up to m/z 100 000 by laser ionization time-of-flight mass spectrometry. Rapid Commun Mass Spectrom 2(8):151–153

13. Lin YS, Chen YC (2002) Laser desorption/ionization time-of-flight mass spectrometry on sol-gel-derived 2,5-dihydroxybenzoic acid film. Anal Chem 74(22):5793–5798

14. Goto-Inoue N, Hayasaka T, Zaima N et al (2010) The detection of glycosphingolipids in brain tissue sections by imaging mass spectrometry using gold nanoparticles. J Am Soc Mass Spectrom 21(11):1940–1943

15. Hayasaka T, Goto-Inoue N, Zaima N et al (2010) Imaging mass spectrometry with silver nanoparticles reveals the distribution of fatty acids in mouse retinal sections. J Am Soc Mass Spectrom 21(8):1446–1454

16. Shrivas K, Hayasaka T, Sugiura Y et al (2011) Method for simultaneous imaging of endogenous low molecular weight metabolites in mouse brain using TiO2 nanoparticles in nanoparticle-assisted laser desorption/ionization-imaging mass spectrometry. Anal Chem 83(19):7283–7289

17. Taira S, Sugiura Y, Moritake S et al (2008) Nanoparticle-assisted laser desorption/ionization based mass imaging with cellular resolution. Anal Chem 80(12):4761–4766

18. Ageta H, Asai S, Sugiura Y et al (2009) Layer-specific sulfatide localization in rat hippocampus middle molecular layer is revealed by nanoparticle-assisted laser desorption/ionization imaging mass spectrometry. Med Mol Morphol 42(1):16–23

19. Chiang CK, Chen WT, Chang HT (2011) Nanoparticle-based mass spectrometry for the analysis of biomolecules. Chem Soc Rev 40(3):1269–1281

20. Armarego WLF, Chai CLL (2003) Purification of laboratory chemicals, 5th edn. Butterworth-Heinemann, Oxford, p 608

21. Chen X, Mao SS (2007) Titanium dioxide nanomaterials: synthesis, properties, modifications, and applications. Chem Rev 107(7):2891–2959

22. Guillou H, Zadravec D, Martin PG et al (2010) The key roles of elongases and desaturases in mammalian fatty acid metabolism: insights from transgenic mice. Prog Lipid Res 49(2):186–199

23. Thies F, Garry JM, Yaqoob P et al (2003) Association of n-3 polyunsaturated fatty acids with stability of atherosclerotic plaques: a randomised controlled trial. Lancet 361(9356):477–485

24. Uauy R, Dangour AD (2006) Nutrition in brain development and aging: role of essential fatty acids. Nutr Rev 64(5 Pt 2):S24–S33, discussion S72-91

25. van Meer G, de Kroon AI (2011) Lipid map of the mammalian cell. J Cell Sci 124(Pt 1):5–8

26. Hayasaka T, Goto-Inoue N, Sugiura Y et al (2008) Matrix-assisted laser desorption/ionization quadrupole ion trap time-of-flight (MALDI-QIT-TOF)-based imaging mass spectrometry reveals a layered distribution of phospholipid molecular species in the mouse retina. Rapid Commun Mass Spectrom 22(21):3415–3426

27. Hannun YA, Bell RM (1989) Functions of sphingolipids and sphingolipid breakdown products in cellular regulation. Science 243(4890):500–507

28. Hakomori S, Handa K, Iwabuchi K et al (1998) New insights in glycosphingolipid function: "glycosignaling domain," a cell surface assembly of glycosphingolipids with signal transducer molecules, involved in cell adhesion coupled with signaling. Glycobiology 8(10):xi–xix

29. Furukawa K, Takamiya K, Okada M et al (2001) Novel functions of complex carbohydrates elucidated by the mutant mice of glycosyltransferase genes. Biochim Biophys Acta 1525(1–2):1–12

30. Sugiura Y, Konishi Y, Zaima N et al (2009) Visualization of the cell-selective distribution of PUFA-containing phosphatidylcholines in mouse brain by imaging mass spectrometry. J Lipid Res 50(9):1776–1788

31. Sugiura Y, Setou M (2010) Imaging mass spectrometry for visualization of drug and endogenous metabolite distribution: toward in situ pharmacometabolomes. J Neuroimmune Pharmacol 5(1):31–43

32. Walch A, Rauser S, Deininger SO et al (2008) MALDI imaging mass spectrometry for direct tissue analysis: a new frontier for molecular histology. Histochem Cell Biol 130(3):421–434

33. Schwartz SA, Reyzer ML, Caprioli RM (2003) Direct tissue analysis using matrix-assisted laser desorption/ionization mass spectrometry: practical aspects of sample preparation. J Mass Spectrom 38(7):699–708

Chapter 17

Matrix-Enhanced Surface-Assisted Laser Desorption/Ionization Mass Spectrometry (ME-SALDI-MS) for Mass Spectrometry Imaging of Small Molecules

Victoria L. Brown, Qiang Liu, and Lin He

Abstract

Surface-assisted laser desorption/ionization mass spectrometry (SALDI-MS), a parallel technique to matrix-assisted laser desorption/ionization mass spectrometry (MALDI-MS), utilizes inorganic particles or porous surfaces to aid in the desorption/ionization of low-molecular-weight (MW) analytes. As a matrix-free and "soft" LDI approach, SALDI offers the benefit of reduced background noise in the low MW range, allowing for easier detection of biologically significant small MW species. Despite the inherent advantages of SALDI-MS, it has not reached comparable sensitivity levels to MALDI-MS. In relation to mass spectrometry imaging (MSI), intense efforts have been made in order to improve sensitivity and versatility of SALDI-MSI. We describe herein a detailed protocol that utilizes a hybrid LDI method, matrix-enhanced SALDI-MS (ME-SALDI MS), to detect and image low MW species in an imaging mode.

Key words Surface-assisted laser desorption/ionization mass spectrometry, Matrix-enhanced SALDI-MS, Mass spectrometry imaging, Ionic matrix, Metabolite

1 Introduction

Developed in parallel to MALDI, surface-assisted laser desorption/ionization (SALDI) offers a matrix-free approach to LDI mass spectrometry. Initial efforts in developing a surface-based method for MS involved using an inorganic matrix composed of 30 nm cobalt particles suspended in glycerol, which resulted in the detection of proteins exceeding masses of 20,000 Da [1, 2]. Many materials have since been tested as viable candidates for SALDI analyses, such as Au, Si, TiO_2, and Zn nanoparticles; carbon nanotubes; graphite; and nanostructured thin metal films [3–9]. Desorption ionization on silicon (DIOS) is one of the more successful SALDI methods introduced in 1999 by the Suizdak group where porous silicon (pSi) is utilized as the energy transfer medium to assist in analyte desorption/ionization. Picomole to high femtomole amounts of intact molecular species have been

Lin He (ed.), *Mass Spectrometry Imaging of Small Molecules*, Methods in Molecular Biology, vol. 1203,
DOI 10.1007/978-1-4939-1357-2_17, © Springer Science+Business Media New York 2015

successfully analyzed in DIOS without the presence of matrix [10]. Detection of a diverse group of compounds such as carbohydrates, small MW drug molecules, glycolipids, peptides, and natural products has been demonstrated—thus exemplifying the versatility of SALDI (DIOS) for MS applications [2, 10].

In SALDI, the elimination of traditional MALDI matrix leads to the removal of low MW background noise, thus making detection of small MW species more facile to accomplish. Despite the absence of organic matrix, however, the unique properties of nanomaterials used in SALDI permit it to maintain the classic MS "soft" ionization characteristics and most analytes are detected intact. As in MALDI, the desorption/ionization process in SALDI is still heavily debated, but extensive research has helped shed light on this dynamic process. Many reports have attributed its effective analyte desorption/ionization to the electronic and thermal properties of nanomaterials of choice [2, 10]. For example, studies have shown rapid heating of the pSi walls as one of the main facilitators of analyte desorption in DIOS [11, 12]. Ionization, on the other hand, is thought to occur through existing surface charges on the pSi surface or through solvent-analyte interactions within the desorption plume where residual solvent molecules play a role as a pseudo-matrix. Additional factors that have been found to play a role in ionization include analyte proton affinity, pore depth, surface roughness, and the likelihood of the substrate to form free electron/hole pairs [12–14].

To further improve the ionization efficiency of pSi-based SALDI method, a hybrid ionization method, i.e., matrix-enhanced SALDI (ME-SALDI), has been developed that combines valuable attributes from both conventional MALDI and SALDI methods [15]. In ME-SALDI, the introduction of a thin layer of conventional MALDI matrix, such as α-cyano-4-hydroxycinnamic acid (CHCA) or 2,5-dihydroxybenzoic acid (DHB), provides a proton-rich environment that enhances ionization efficiency of desorbed species. At the same time, the presence of pSi substrates reduces the laser fluence needed for analyte desorption; subsequently few matrix molecules co-desorb and less background noise is observed in resulting spectra. ME-SALDI MSI has shown substantially improved MS performance over conventional MALDI or SALDI methods with reduced matrix interference and analyte fragmentation, larger mass detection window, and much improved analyte ionization efficiency. The use of ionic matrix, a conventional MALDI matrix paired with an organic base for better vacuum stability, further improves matrix deposition homogeneity and reduces matrix background interference [16, 17].

Here we describe an analytical protocol for ME-SALDI MS imaging using either traditional MALDI matrices or ionic matrices, CHCA/Py (α-cyano-4-hydroxycinnamic acid:pyridine) and CHCA/ANI (α-cyano-4-hydroxycinnamic acid:aniline) atop a pSi substrate. Matrix synthesis, sample preparation, and optimization of experimental conditions that are critical to a successful ME-SALDI MSI experiment are described.

2 Materials

2.1 Preparation of Porous Silicon (pSi) Substrates

1. N-type Sb-doped (100) single-crystalline silicon wafers of low resistivity, e.g., 0.005–0.02 Ω/cm.
2. pSi etching solution: 1:1 (v/v) of 49 % hydrofluoric acid (HF):95 % ethanol.
3. pSi wash solution: 1:10 (v/v) of 49 % HF:95 % ethanol.
4. pSi oxidation solution: 1:1 (v/v) 30 % hydrogen peroxide (H_2O_2):95 % ethanol.

2.2 Tissue Section Preparation

1. Optical cutting temperature (OCT) compound.
2. Methylene blue staining solution: Dissolve 0.15 g methylene blue powder in 100 mL of 70 % ethanol. Stir solution overnight.
3. Leica Stainless-Steel Re-usable Microtome Knives.
4. Carbon steel surgical blades.
5. Plant tissue for metabolite imaging, such as garlic samples used in this protocol.

2.3 Matrix Deposition

1. 2,5-Dihydroxybenzoic acid (DHB) and α-cyano-4-hydroxy-cinnamic acid (CHCA).
2. Solid ionic matrixes: CHCA/pyridine (CHCA/Py) and CHCA/aniline (CHCA/ANI) (see Subheading 3.2).
3. Silicon oil.
4. Sublimation glassware (see **Note 1**).
5. Double-sided conductive tape.

2.4 Instrumentation

1. A potentiostat, such as the one by EG&G Princeton, Model 273.
2. A vacuum pump equipped with a vacuum meter.
3. A rotovap.
4. A cryo-cut microtome, such as the one by American Optical Corp., Buffalo, NY, USA.
5. An optical microscope, such as Leica DMRX equipped with a Donpisha XC-003P CCD camera.
6. Applied Biosystems Voyager DE-STR MALDI-TOF mass spectrometer (Framingham, MA).
7. MMSIT MALDI Imaging Tool software V2.2.0 (©2004 by Markus Stoeckli, Novartis Institutes for BioMedical Research,

Basel, Switzerland, newer version free downloadable at www.
maldi-msi.org).

8. BioMap 3.7.5.4 (by Markus Stoeckli, Novartis Institutes for
   BioMedical Research, Basel, Switzerland, newer version free
   downloadable at www.maldi-msi.org).

3 Methods

3.1 Preparation of Porous Silicon Substrate

1. Cut silicon wafers into 1-cm² square-shaped chips and dip into
   the pSi wash solution for 1 min to remove the oxidized layer
   (caution: HF can cause severe tissue damage upon contact or
   inhalation). After removing the oxidized layer, wash the Si
   chips in 95 % ethanol and dry with N_2.

2. Assemble a Si chip in a Teflon anodic etching cell, as shown in
   Fig. 1. The etching cell consists of a three-electrode system
   where a Au working electrode is placed under the Si chip, and a
   Pt counter electrode and a Pt reference electrode are placed
   above the surface. Clamp the cell tightly prior to filling it with
   the pSi etching solution. Align a 59-W tungsten lamp through a
   concave lens to provide uniform illumination to the Si surface.
   Program a potentiostat for desired etching current and the time
   (example parameters given in the following step) (*see* **Note 2**).

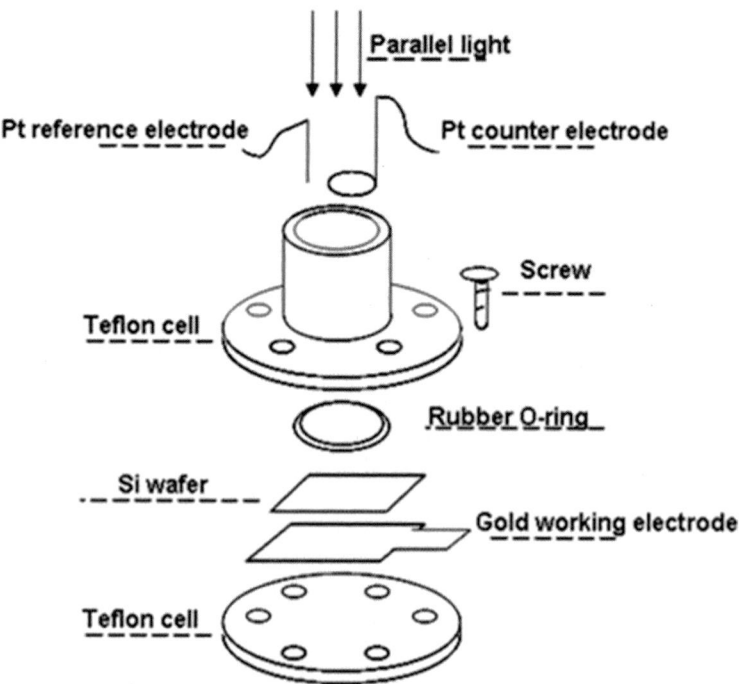

Fig. 1 Schematic draw of an electrochemical etching cell for SALDI substrate
preparation. Adapted with permission from ref. 12

3. Electrochemically etch the Si chip in the pSi etching solution for 1 min at a current density of 5 mA/cm^2. After etching, carefully transfer the HF waste to a pre-labeled waste bottle in the hood dedicated to HF. Wash the setup with 95 % ethanol thoroughly before disassembly. Wash the produced pSi substrate again with copious amounts of 95 % ethanol and dry the substrate under N$_2$.

4. To double-etch the pSi substrate, first dip the substrate in the pSi oxidation solution for 1 min, followed by washing in 95 % ethanol, drying in a N$_2$ stream, and dipping in the pSi wash solution for 1 min. Wash the substrate again in 95 % ethanol and store in ethanol until needed (*see* **Note 3**).

5. Prior to usage, refresh the substrates in the pSi wash solution for 1 min again, wash in 95 % ethanol, and dry with N$_2$.

3.2 Synthesis of Solid Ionic Matrix

1. Dissolve 38.5 mg of CHCA in 20 mL of methanol in a round-bottom flask. Add 16.6 mL of Py or 18.7 mL of ANI into the CHCA solution.

2. Stir the reaction mixture for 2 h at room temperature.

3. Remove the solvent by rotovap and dry the yielded solid ionic matrix under vacuum.

4. Store CHCA/ANI under vacuum at room temperature before use.

3.3 Tissue Section Preparation

1. Precut fresh garlic glove tissues into thin slices using carbon steel surgical blades with at least one dimension no less than 2 mm. Snap freeze all thin tissue slices in liquid N$_2$ immediately and store at –80 °C prior to usage (*see* **Note 4**).

2. Use a cryo-cut microtome equipped with stainless-steel microtome knives for tissue sectioning. Procedural modification may be needed when different equipment is being used. Mount the frozen tissue slice onto the microtome chunk, using the OCT compound as a "glue" to hold to tissue in place. Slowly bring the temperature up to –20 °C. Section the frozen tissue samples into 10-μm-thick slices. Transfer the 10-μm-thick tissue slices onto a dry pSi substrate and slowly bring it up to room temperature.

3. Meanwhile, transfer the adjacent section of the same tissue chunk onto a microscope glass slide for conventional histological staining. Deposit the methylene blue staining solution (0.15 %, 200 μL) onto the tissue section. After 10 s of staining, wash the excess staining solution away with 95 % ethanol.

4. Use an optical microscope equipped with a digital camera to collect the optical images of stained tissue sections to provide visual validation of MSI data.

3.4 Matrix Deposition

1. Set up the sublimation apparatus in a chemical hood (Fig. 2). Use a vacuum pump with a vacuum meter to provide controllable vacuum in the sublimation chamber.

2. Add 0.3 g of CHCA, DHB, CHCA/Py, or CHCA/ANI to the bottom of the sublimation chamber.

3. Attach the tissue-coated porous Si to the flat bottom of the apparatus condenser, facing downwards to the matrix.

4. Cool the condenser, along with the porous silicon substrate, with running water for vapor condensation (*see* **Note 5**).

5. Preheat the silicon oil bath to 110 °C for DHB, to 113 °C for CHCA/Py, to 120 °C for CHCA, or to 170 °C for CHCA/ANI. Maintain the chamber pressure at ~50 Torr for 2 min before immersing the sublimation glassware into oil bath.

6. For CHCA or CHCA/Py, heat the sublimation setup for 5 min. Use a 10-min heating time for CHCA/ANI sublimation and a 1–5-min heating time for DHB sublimation (*see* **Note 6**).

7. Remove the sublimation setup from the oil bath and release the vacuum gently when the sublimation chamber cools down to the room temperature. Take the pSi substrate coated with matrix out of the sublimation apparatus and immediately load it into the MS sample chamber for MS analysis. Turn off the vacuum pump, the cooling water, and the hot plate. Excess matrix left in the sublimation chamber can be saved in a separate bottle for future usage but not to be put back to the original bottle to avoid contamination.

Fig. 2 A photo picture of in-house sublimation apparatus. Adapted with permission from ref. 12

3.5 Mass Spectrometry Imaging

Modification in instrumental conditions may be needed to achieve optimal imaging results when different instruments are used.

1. The mass spectrometer used in this protocol is equipped with a 20-Hz N_2 laser. Adjust the laser irradiation energy by a neutral density filter and the beam size to 35 μm with an adjustable pinhole placed close to the laser entrance window. The actual laser beam size can be measured by increasing laser fluence till a burn mark is left behind on a SALDI substrate to allow off-line measurements of the laser beam footprint (*see* **Note 7**).

2. Mount the matrix-coated pSi substrate with a piece of tissue section attached onto a stainless steel MALDI plate with a piece of double-sided conductive tape.

3. Operate the instrument at an accelerating voltage of 20 kV in the reflector and positive ion mode. For the data presented here, 50 laser shots were accumulated at each imaging pixel. The translational stage was operated at 50-μm stepwise.

4. The MS instrument is controlled by MMSIT MALDI Imaging Tool software V2.2.0. The imaging area is manually selected along the outline of the tissue section in MMSIT.

5. 2-D ion maps are reconstructed using BioMap 3.7.5.4. An example of the imaging results is shown in Fig. 3.

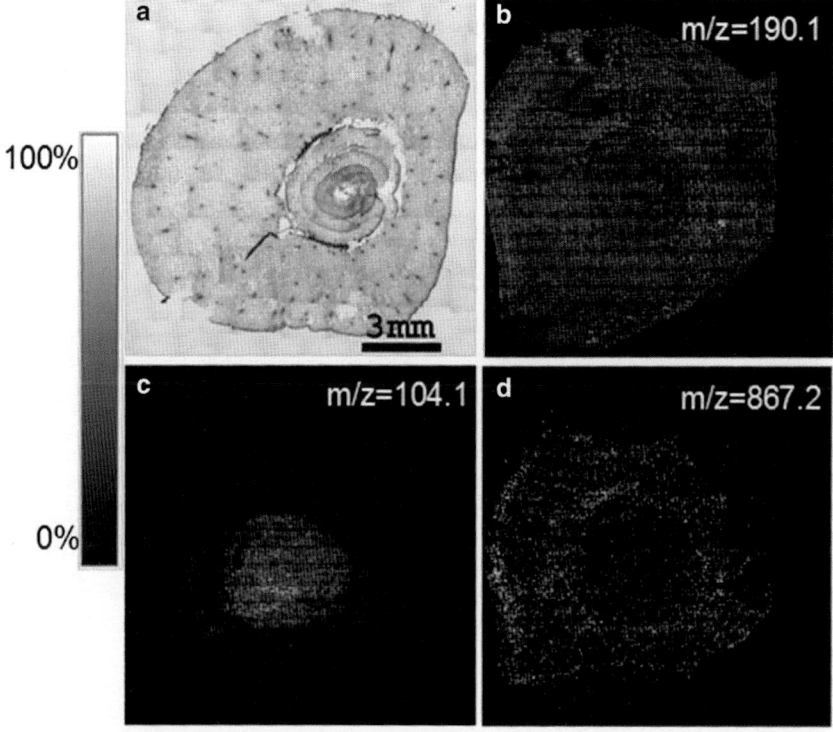

Fig. 3 (**a**) An optical image of 10-μm-thick garlic section. (**b**) Reconstructed 2-D images for ions at (**c**) m/z=190.1, (**d**) m/z=104.1, and (**e**) m/z=867.2 from CHCA/ANI-coated garlic section in ME-SALDI IMS. Adapted with permission from ref. 17

4   Notes

1. Figure 2 shows a typical glassware setup used in matrix deposition setup. Different designs can be used as long as the chamber is sealed under vacuum and the substrate is placed under a lower temperature than the matrix to allow vapor deposition upon contact.

2. During the etching process, gas bubbles are mildly, but continuously released. The Si chip should turn from grey to bright blue first, and then quickly change to a golden color. Over time, the color of the Si chip will become darker and darker, suggesting formation of a rough surface. The final pSi substrate should exhibit a dark blue hue after dried. Formation of porous features on Si is critically related to the property of Si wafers, the amount of irradiating light, the current density applied, and the etching time; hence it should be optimized for individual etching setup and Si substrates. The color of the resulting pSi substrate can be used as a reference point in troubleshooting. For example, a substrate with a yellowish color suggests over-etching whereas a light grey hue suggests under-etching of the substrate. It is important to note that the complete etching setup should be placed in a chemical hood and all operations prior to pSi substrate drying under the N_2 stream should be carried out in a chemical hood, with proper safety measures taken for HF handling.

3. Severe MS performance degradation has been correlated to exposure of porous Si to air. The pSi substrates are therefore required to be stored in ethanol until needed. For the substrates coated with tissue sample, matrix deposition should be carried out immediately out of the same concern.

4. Freeze and thaw of tissue samples is a critical step in preparation of tissue sections. The most common problem observed in frozen tissue sections is the ice crystal damage, which causes leaky tissues and blotched tissue surface. The size of ice crystals is usually determined by the speed of the freezing process through the whole tissue. Therefore, a pre-cutting step is recommended to reduce the overall tissue size before snap freezing.

5. Use a circulating water system under the room temperature to avoid moisture condensation atop the surface of porous silicon substrate.

6. Sublimation is sensitive to the vacuum pressure and the temperature in the chamber. The conditions described here are optimized for our setup and the deposition efficiency may vary among laboratories. Direct visual inspection of matrix formation is recommended by placing the resulting substrate under an optical microscope. An amorphous thin layer of matrices

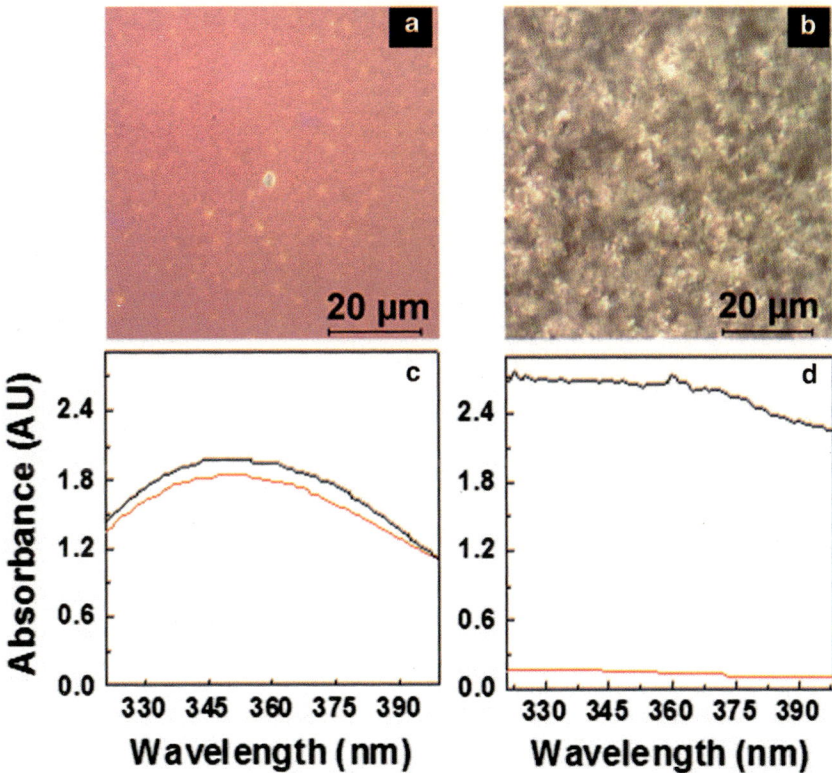

Fig. 4 Optical images of (**a**) the solid ionic matrix (CHCA/ANI) and (**b**) conventional matrix (DHB) sublimated on transparent glass slides. Representative UV–vis spectra of (**c**) the CHCA/ANI layer and (**d**) the DHB layer collected before (*black line*) and after (*red line*) 40-h vacuum storage. Adapted with permission from ref. 17

without apparent crystal formation is preferred (Fig. 4). To quantitatively control the amount and the quality of the matrix deposited, a clear glass side should be coated under each sublimating condition in parallel. Empirically it has been found that a matrix layer with UV absorbance between 1.0 and 2.0 is optimal for ME-SALDI.

7. Adjust the irradiation energy to achieve optimal MS performance prior to sample imaging. Irradiation energies for SALDI-MS are typically significantly lower than the energy needed for traditional MALDI.

References

1. Shanta SR et al (2011) Binary matrix for MALDI imaging mass spectrometry of phospholipids in both ion. Anal Chem 83:1252

2. Ayorinde FO et al (1999) Use of meso-tetrakis(pentafluorophenyl)porphyrin as a matrix for low molecular weight alkylphenol ethoxylates in laser desorption/ionization

time-of-flight mass spectrometry. Rapid Commun Mass Spectrom 13:2474–2479

3. Knochenmuss R (2006) Ion formation mechanisms in UV-MALDI. Analyst 131:966–986

4. Peterson DS (2007) Matrix-free methods for laser desorption/ionization mass spectrometry. Mass Spectrom Rev 26:19–34

5. Nayak R et al (2010) Matrix-free LDI mass spectrometry platform using patterned nano-structured gold thin film. Anal Chem 82: 7772–7778

6. Watanabe T et al (2008) Surface-assisted laser desorption/ionization mass spectrometry (SALDI-MS) of low molecular weight organic compounds and synthetic polymers using zinc oxide (ZnO) nanoparticles. J Mass Spectrom 43:1063–1071

7. Sunner J et al (1995) Graphite surface-assisted laser desorption/ionization time-of-flight mass spectrometry of peptides and proteins from liquid solutions. Anal Chem 67:4335–4342

8. Kraft P et al (1998) Infrared, surface-assisted laser desorption ionization mass spectrometry on frozen aqueous solutions of proteins and peptides using suspensions of organic solids. J Am Soc Mass Spectrom 9:912–924

9. Xu S et al (2003) Carbon nanotubes as assisted matrix for laser desorption/ionization time-of-flight mass spectrometry. Anal Chem 75: 6191–6195

10. Arakawa R et al (2010) Functionalized nanoparticles and nanostructured surfaces for surface-assisted laser desorption/ionization mass spectrometry. Anal Sci 26:1229–1240

11. Wei J et al (1999) Desorption-ionization mass spectrometry on porous silicon. Nature 399:243–246

12. Liu Q et al (2010) Mass Spectrometry Imaging of Small Molecules Using Matrix-Enhanced Surface-Assisted Laser Desorption/Ionization Mass Spectrometry (SALDI-MS) In: Mass Spectrometric Imaging: History, Fundamentals and Protocols, Methods in Molecular Biology Series, Sweedler JV, Rubakhin, S. Eds. Humana. vol 656, pg 243

13. Finkel NH (2005) Surface-assisted laser desorption/ionization-mass spectrometry (SALDI-MS) of controlled nanopore cavities and the associated thermal properties. North Carolina State University, Raleigh

14. Sailor MJ (2011) Fundamentals of porous silicon preparation. In: Porous silicon in practice: preparation, characterization and applications. Wiley-VCH Verlag GmbH & Co. KGaA, p 1–42. doi: 10.1002/9783527641901.ch1

15. Lui Q et al (2009) Metabolite imaging using matrix-enhanced surface-assisted laser desorption/ionization mass spectrometry (ME-SALDI-MS). J Am Soc Mass Spectrom 20:80–88

16. Armstrong DW et al (2001) Ionic liquids as matrixes for matrix-assisted laser desorption/ionization mass spectrometry. Anal Chem 73:3679–3686

17. Lui Q et al (2009) Ionic matrix for surface-assisted laser desorption ionization mass spectrometry imaging (ME-SALDI-MSI). J Am Soc Mass Spectrom 20:2229–2237

Chapter 18

Laser Desorption Postionization Mass Spectrometry Imaging of Biological Targets

Artem Akhmetov, Chhavi Bhardwaj, and Luke Hanley

Abstract

Laser desorption photoionization mass spectrometry (LDPI-MS) utilizes two separate light sources for desorption and photoionization of species from a solid surface. This technique has been applied to study a wide variety of molecular analytes in biological systems, but is not yet available in commercial instruments. For this reason, a generalized protocol is presented here for the use of LDPI-MS imaging to detect small molecules within intact biological samples. Examples are provided here for LDPI-MS imaging of an antibiotic within a tooth root canal and a metabolite within a coculture bacterial biofilm.

Key words Laser desorption, Vacuum ultraviolet, Postionization, Single photon ionization, Mass spectrometry, Imaging, Biofilms, Tooth

1 Introduction

Laser desorption photoionization mass spectrometry (LDPI-MS) utilizes two separate light sources for desorption and photoionization of species from a solid surface [1, 2]. Briefly, the first pulsed laser desorbs a plume of neutral molecules from the sample surface, followed by a vacuum ultraviolet (VUV) laser or other light source that intercepts the desorbed plume and single photon ionizes (SPI) the gaseous neutrals [3]. The resulting ions are then detected by time-of-flight or other mass analyzer. In order for a desorbed molecule or cluster to ionize, the photon energy of the ionizing light source must be higher than its ionization energy. Selection of the photon energy thereby allows some selectivity in the analysis process [2, 4, 5]. The desorption laser fluence is kept low to minimize direct ionization and fragmentation. Laser desorption followed by photoionization detects abundant neutrals that can greatly improve limits of detection, allowing for measurement of compounds with potentially better sensitivity and greater analyte information compared to matrix-assisted laser desorption ionization and other single-laser desorption/ionization methods. Separation of

Lin He (ed.), *Mass Spectrometry Imaging of Small Molecules*, Methods in Molecular Biology, vol. 1203,
DOI 10.1007/978-1-4939-1357-2_18, © Springer Science+Business Media New York 2015

desorption and ionization also facilitates accurate quantitation [6]. This technique has been applied to study various molecular analytes in biological systems including antibiotics, peptides, and metabolites in bacterial biofilms [2, 7, 8]; cholesterol in animal tissue [9]; and endogenous species in soil organic matter [10].

There are currently no commercial LDPI-MS instruments, but there are several custom instruments built in the last dozen or so years which are listed here. References to instruments built before the year 2000 can be found elsewhere [11]. The Argonne National Laboratory group described an LDPI-MS imaging instrument which builds on that laboratory's long tradition in this area [12] and which has been used for experiments similar to those described below [13]. The authors' group at the University of Illinois at Chicago has described two instruments, both inspired by the Argonne instrument [2, 8, 14]. The Ahmed group at the Lawrence Berkeley National Laboratory has modified a commercial time-of-flight secondary ion mass spectrometer to allow imaging with a desorption laser and achieves SPI via VUV radiation from a synchrotron [7, 15]. The Zare group at Stanford University has also described an LDPI-MS instrument which uses far-IR laser desorption, which they refer to as a two-laser mass spectrometer or L2MS [16]. The He group of the National Synchrotron Radiation Laboratory in Hefei, China, has described a custom LDPI-MS using mid-IR desorption and synchrotron radiation for postionization [17]. The Zare and other groups have also reported on the use of multiphoton ionization instead of SPI in LDPI-MS [3, 18–20]. All of the instruments reported above use either linear or reflectron time-of-flight mass analyzers, but they differ considerably in the design of their ion sources and not all allow MS imaging.

Given the absence of commercial instruments, the following protocol has been generalized for use with any LDPI-MS imaging instrument that utilizes VUV lasers for SPI. Prior work has discussed important factors in analysis [2, 5, 18]. Examples are provided here for the protocols for analysis of two biological samples: a bacterial biofilm and a tooth root canal doped with an antibiotic.

2   Description of Typical LDPI-MS Instrument

A detailed description of the custom-built instrument used for the experiments performed here is reported elsewhere and is summarized in Fig. 1 [2, 8]. Briefly, the instrument utilizes a 349 nm Nd:YLF desorption laser, typically running at 25–35 µJ desorption energy with 20–50 µm diameter spot size. The instrument is equipped with two ionization lasers, a 157 nm fluorine excimer laser which generates 7.87 eV photons, and a 355 nm Nd:YAG

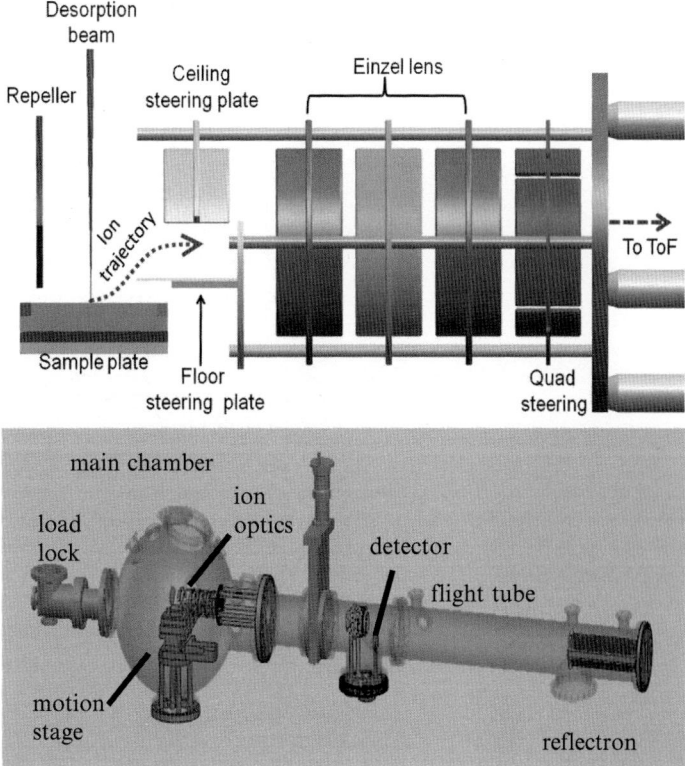

Fig. 1 (*Top*) Ion source and (*bottom*) vacuum chamber for laser desorption postionization mass spectrometer (LDPI-MS) instrument. Ion source shows only desorption laser, as ionization lasers come out of page parallel to and immediately above surface of sample plate. Load lock, main chamber, sample holder with motion stage, ion optics, reflectron, detector, and time-of-flight (ToF) tube are shown in vacuum chamber drawing

laser which generates 10.5 eV photons by tripling in Xe gas, as described previously [8, 14]. The 10.5 eV photon energy is of sufficient energy to ionize a wide range of compound classes, while the 7.87 eV allows for detection of a select class of compounds such as polyaromatic hydrocarbons and some pharmaceuticals that have ionization energies below 7.87 eV. The ionization lasers are typically operated at 100 Hz for the 7.87 eV ionization laser, and 10 Hz for the 10.5 eV ionization laser. The ion source for LDPI-MS has recently evolved through several generations [2, 8, 12, 14], with the second generation source depicted in Fig. 1 [2].

3   Instrument Workflow for Experiment

Figure 2 shows an overview of the workflow of each experiment involving LDPI-MS imaging analysis, emphasizing sample introduction since such customized instruments lack the automated

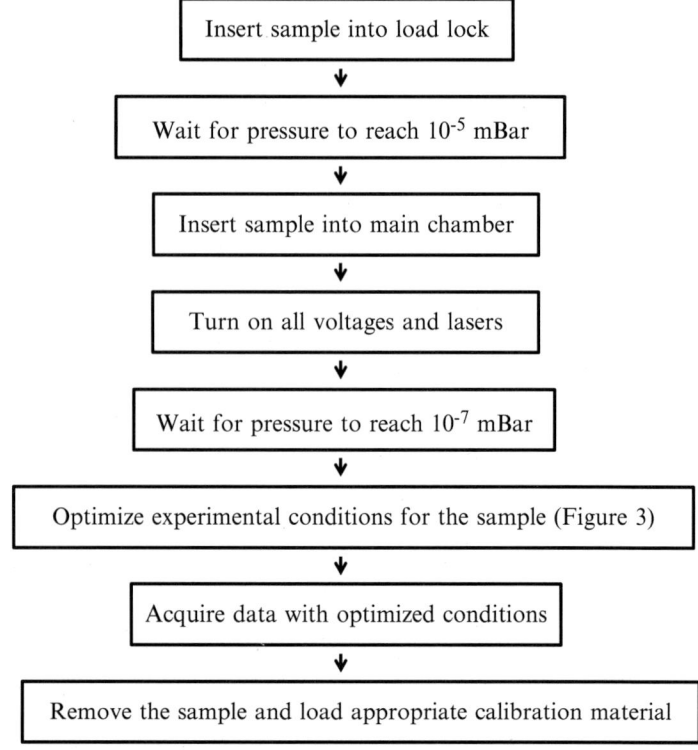

Fig. 2 Instrument workflow for each experiment. The sample is prepared and inserted into the instrument followed by data acquisition

sample transfer stages that are standard in most commercial MS instruments. The high-vacuum conditions required to operate the time-of-flight mass analyzer require that any residual moisture be removed from the sample by leaving it under vacuum in the load lock until the pressure reaches $\sim 10^{-5}$ mBar. Next, the sample is inserted into the main chamber and is allowed to reach an equilibrium pressure of $\sim 10^{-7}$ mBar. Power supplies for ion optics and the detector are turned on at this time and allowed to stabilize for >15 min. Analysis of the sample can begin after pressure drops below 10^{-7} mBar. Acquisition parameters and ion optics should have already been optimized prior to final data acquisition (*see* below). Since the values of these parameters affect the mass accuracy of the instrument, the acquired data must usually be reanalyzed using a mass calibration recorded on the same day.

4   Optimization of Data Acquisition Parameters

Optimization of data acquisition parameters is shown in Fig. 3. Data acquisition begins with acceleration and ion lens voltages initially determined by an ion optical calculation, and then refined

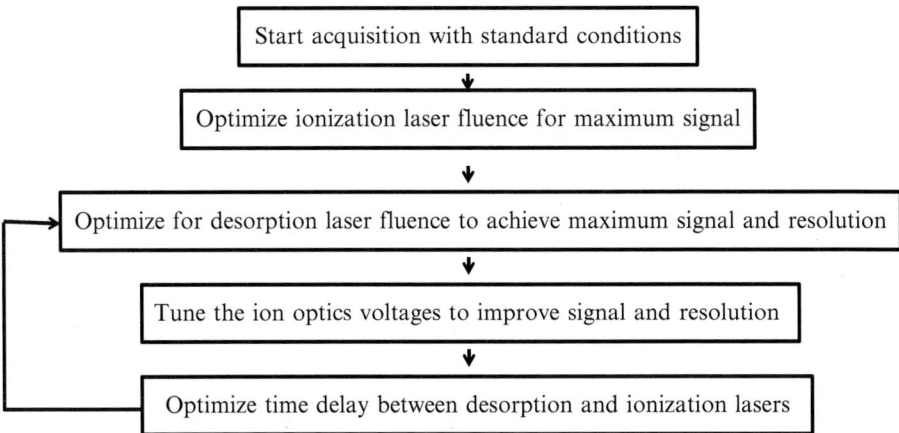

Fig. 3 Optimization of data acquisition parameters. The acquisition begins with standard conditions, and then ion optic voltages are optimized for optimum signal and resolution

through subsequent experiments. This set of standard conditions is stored on the data acquisition computer to be utilized as the initial instrumental settings at the start of each daily experimental run. The default delay between desorption and ionization lasers is set on the time delay generator box, but is optimized for different experiments.

When the procedure in Fig. 2 is completed up to the actual collection of MS data, the standard conditions are loaded and a series of daily optimizations are performed. First, ionization laser fluence is optimized for maximum signal. Two different ionization lasers are available, but the procedure varies slightly with each laser. In both cases, laser optics along the VUV beam path must be occasionally cleaned to maintain maximum ionization laser fluence [21].

For the 7.87 eV fluorine laser, the laser fluence is monitored at all times during the experiment. The laser may be refilled with fluorine gas if fluence is too low to obtain optimum signal intensity, as verified by collecting SPI-MS of triethylamine vapor introduced into the vacuum chamber via a leak valve. Next, the desorption laser is turned on and LDPI-MS signal is optimized on known standard compounds spotted onto a sample plate. Long data acquisition times may require refills if the laser fluence drops excessively during the experiment.

LDPI-MS with 10.5 eV radiation requires a Xe or Xe:Ar gas cell used for tripling the 355 nm output of a Nd:YAG [8, 14]. The Nd:YAG fluence at 355 nm is optimized for maximum 10.5 eV output followed by optimization of Xe gas pressure. If a Xe:Ar cell is used, then the gas ratio must also be optimized. 10.5 eV generation and output are measured by SPI-MS of gaseous acetone introduced via leak value. However, the absolute gas pressure, gas ratios (if relevant), and Nd:YAG pump fluence tend to remain constant over many experiments. While 10.5 eV generation has not been always reliable in some experimental configurations [1], this difficulty has been addressed through careful attention to the details of gas cell design, cleanliness of the gas cell [8, 14], and quality of the LiF optics used to transmit and focus VUV radiation [21].

Once ionization laser fluence is maximized, desorption laser fluence must be optimized during each experiment to maximize molecular ion signal while minimizing ion fragmentation. Increasing desorption laser fluence increases total ion signal, but this can be accompanied by an increase in the total fragmentation (up to a certain threshold) due to enhanced transfer of internal energy into the desorbed molecular species. However, it has been reported that surpassing this desorption laser fluence threshold can lead to reduced fragmentation for some organic samples [5]. Thus, the goal of the optimization of the desorption laser fluence is to minimize the fragmentation of the molecular ion while maximizing the total signal for that ion. Furthermore, optimization of desorption laser fluence is usually dependent on the type of sample and the nature of the experiment. For example, biofilm samples typically require higher fluences in order to desorb the necessary amount of sample for successful analysis.

Finally, laser spot size must be optimized for most experiments. Typically, the goal is to minimize the total spot size for the highest spatial resolution in imaging mode. However, the laser may be defocused in order to desorb a larger amount of analyte for the samples near the limit of detection. The desorption laser is focused onto the sample by a movable lens which can adjust the focus of the beam on the sample. Varying the position of the lens allows for varying the laser spot size on the sample. The optimum position of the lens and the resulting spot size vary from experiment to experiment, with larger spot sizes typically resulting in an increase of total signal.

Various ion optics are used in order to guide the ion plume to the detector. The voltages are tuned to optimize the trajectory of photoions to the detector for maximum signal. Often, voltages are tuned for each position on the sample plate. Each position on the acquisition plate requires voltage fine-tuning corresponding to the electric field. Each sample requires slightly different voltage tuning, since samples can vary in their response to the electric field generated by the desorption plate and thus require adjustment of steering plates for successful detection. Steering and repeller plate voltages are optimized for each experiment to achieve maximum signal at highest resolution.

During each experiment, the time delay between desorption and ionization lasers also must be optimized. Selecting the delay between desorption and ionization laser allows for a degree of control over the fragmentation of the analyte, with longer delays leading to reduced fragmentation and often detection of a higher mass distribution [5]. However, selection of especially long delay times comes at the price of decreased total signal.

Optimization of the desorption laser fluence, followed by the tuning of the voltages and optimization of the time delay between desorption and ionization laser, is repeated until acceptable mass resolution and signal to noise is achieved. However, this optimization can alter the mass calibration of the instrument, as described above, since varying ion optical voltages affects the mass calibration. For this reason, the mass analyzer needs to be calibrated for several different regions of the sample plate to obtain maximum mass resolution.

5   Data Collection for Imaging

Several additional conditions need to be established for imaging experiments. The instrument shown in Fig. 1 employs fixed laser positions with respect to the ion source and ToF mass analyzer, requiring sample-stage translation for imaging. Meandering speed for the sample stage is determined in accordance with the desired resolution, step size, and desired length of acquisition time for the experiment: a typical stage speed is 0.05 mm/s with a step size of 50–100 μm. The operator also controls the total binning and averaging of spectra recorded from individual laser shots. The larger average values result in better signal to noise for the analyte, in results in the trade-off of decreased spatial resolution. The total time required to acquire a mass spectrum is determined by the signal to noise and the data acquisition rate, the latter of which is usually limited by the repetition rate of the ionization laser. Maximum data acquisition rates of 100 Hz for 7.87 eV photoionization and 10 Hz for 10.5 eV photoionization are possible for the instrument described here.

The final step in acquisition is the import of the collected data file into an imaging software program (BioMap, downloaded from www.maldi-msi.org). The final data file is converted to a compatible format utilizing a custom-written software, and imported into the imaging software (BioMap) for generation of the desired mass spectral image.

6   Antibiotic Image on Root Canal of Tooth

MS imaging of thick and/or insulating substrates is often complicated by the distortion of ion desorption that arises from sample charging, but the detection of desorbed neutrals by LDPI-MS reduces such charging effects. Figure 4 shows a comparison between the optical and MS image of a bovine tooth treated with an antibiotic. The root canal of the tooth was irrigated with doxycycline containing disinfectant solution, then the irrigated tooth was split into two and mounted on a sample plate, and the exposed

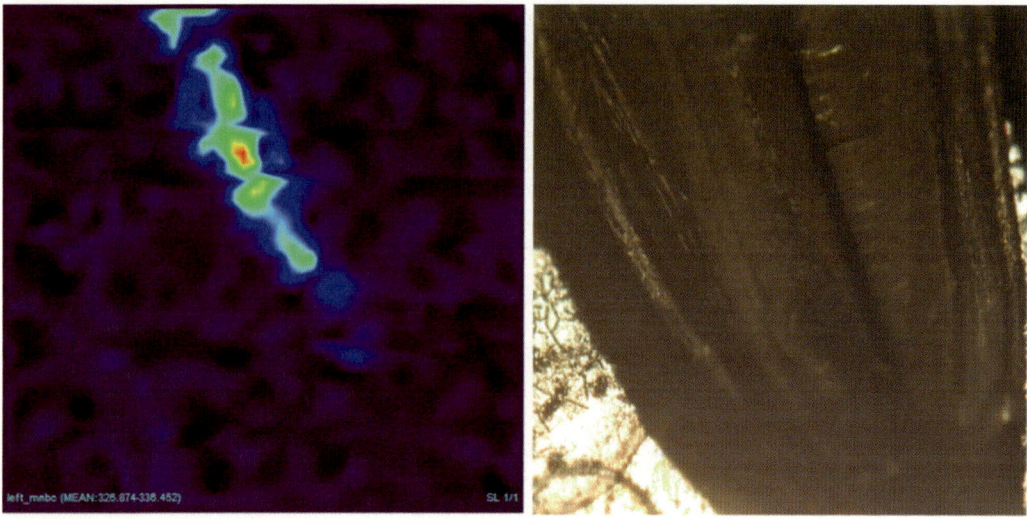

Fig. 4 (*Left*) 7.87 eV LDPI-MS and (*right*) optical images of split tooth after treatment with the antibiotic doxycycline. The *m/z* 326–336 MS image shows high ion intensity for doxycycline along the root canal observed in the optical image (*brighter colors* correspond to higher ion intensities). The images are ~1 cm × 1 cm

root canal and adjacent region were analyzed by LDPI-MS. The MS image can be correlated with the visual image and the penetration of the doxycycline through the root canal can be seen.

7   Metabolite Image on Bacteria Biofilm

Mass spectrometric imaging is an important tool that can be applied towards understanding the biology of intact bacterial biofilms [2, 7, 8]. Coculture *Escherichia coli* bacterial biofilms were imaged using 7.87 eV LDPI-MS. Endogenous metabolites were detected and low mass peaks were obtained. Even though the sample was on an insulating substrate, it was easily analyzed. Figure 5 shows the comparison of the visual image and the extracted ion chromatogram for metabolite signal at *m/z* 42–44. The two *E. coli* biofilms were grown on insulating polycarbonate membranes for analysis. The membranes were then adhered to a stainless steel plate with copper tape for introduction into vacuum for MS analysis. Despite the difficult sample matrix and the insulating sample, the optical and MS images can be correlated.

Biofilm growth procedure and sample blotting have been described previously [8] and are only summarized here. Biofilms were grown on polycarbonate membranes placed on agar. Bacteria were grown in M9 minimal media with 2 g/L sodium acetate and 100 µg/mL ampicillin and citrine was grown in M9 minimal media with 10 g/L glucose and 100 µg/mL ampicillin. The cultures were grown for ~12 h at 37 °C. Membranes were aseptically placed onto

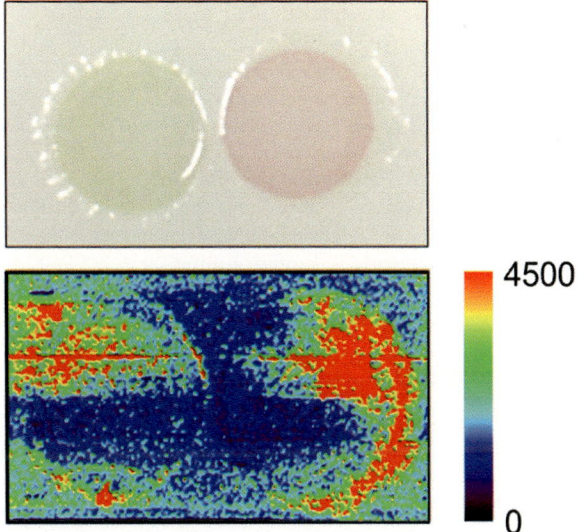

Fig. 5 (*Top*) Optical and (*bottom*) 7.87 eV LDPI-MS image for a two strain *Escherichia coli* coculture biofilms grown on separate 2.5 cm diameter, circular polycarbonate membranes. The optical image shows two distinct biofilms while MS image of *m/z* 42–44 delineates a distribution of a metabolite across different areas of the two biofilms

separate agar plates (three per plate) and each membrane was inoculated with 50 μL of the diluted exponential culture. Inoculated membranes were allowed a short drying period in a laminar hood prior to incubation. The biofilms were grown for ~96–120 h at 37 °C. Membranes were transferred to new plates every 24 h for several days until biofilm growth was sufficient to permit MS analysis.

Acknowledgment

Funding for this project was provided by the National Institutes of Health under grant EB006532 and the University of Illinois at Chicago. The authors acknowledge Jerry F. Moore, Yang Cui, Ross Carlson, and Berdan Sevinc for assistance in various aspects of the experiments.

References

1. Hanley L, Zimmermann R (2009) Light and molecular ions: the emergence of vacuum UV single-photon ionization in MS. Anal Chem 81:4174–4182

2. Akhmetov A, Moore JF, Gasper GL, Koin PJ, Hanley L (2010) Laser desorption postionization for imaging MS of biological material. J Mass Spectrom 45:137–145

3. Zimmermann R (2013) Photo ionisation in mass spectrometry: light, selectivity and molecular ions. Anal Bioanal Chem 405: 6901–6905

4. Blaze MTM, Takahashi LK, Zhou J, Ahmed M, Gasper GL, Pleticha FD, Hanley L (2011) Brominated tyrosine and polyelectrolyte multilayer analysis by laser desorption VUV

postionization and secondary ion mass spectrometry. Anal Chem 83:4962–4969

5. Kostko O, Takahashi LK, Ahmed M (2011) Desorption dynamics, internal energies and imaging of organic molecules from surfaces with laser desorption and vacuum ultraviolet (VUV) photoionization. Chem Asian J 6:3066–3076

6. Blaze MTM, Akhmetov A, Aydin B, Edirisinghe PD, Uygur G, Hanley L (2012) Quantification of antibiotic in biofilm-inhibiting multilayers by 7.87 eV laser desorption postionization MS imaging. Anal Chem 84: 9410–9415

7. Gasper GL, Takahashi LK, Zhou J, Ahmed M, Moore JF, Hanley L (2010) Laser desorption postionization mass spectrometry of antibiotic-treated bacterial biofilms using tunable vacuum ultraviolet radiation. Anal Chem 82: 7472–7478

8. Bhardwaj C, Moore JF, Cui Y, Gasper GL, Bernstein HC, Carlson RP, Hanley L (2013) Laser desorption VUV postionization MS imaging of a cocultured biofilm. Anal Bioanal Chem 405:6969–6977

9. Milasinovic S, Liu Y, Bhardwaj C, Blaze MTM, Gordon RJ, Hanley L (2012) Feasibility of depth profiling of animal tissue by ultrashort pulse laser ablation. Anal Chem 84:3945–3951

10. Liu SY, Kleber M, Takahashi LK, Nico P, Keiluweit M, Ahmed M (2013) Synchrotron-based mass spectrometry to investigate the molecular properties of mineral–organic associations. Anal Chem 85:6100–6106

11. Hanley L, Kornienko O, Ada ET, Fuoco E, Trevor JL (1999) Surface mass spectrometry of molecular species. J Mass Spectrom 34: 705–723

12. Veryovkin IV, Calaway WF, Moore JF, Pellin MJ, Lewellen JW, Li Y, Milton SV, King BV, Petravic M (2004) A new time-of-flight instrument for quantitative surface analysis. Nucl Instrum Meth Phys Res B 219–220:473–479

13. Edirisinghe PD, Moore JF, Skinner-Nemec KA, Lindberg C, Giometti CS, Veryovkin IV,

Hunt JE, Pellin MJ, Hanley L (2007) Detection of in-situ derivatized peptides in microbial biofilms by laser desorption 7.87 eV postionization mass spectrometry. Anal Chem 79:508–514

14. Cui Y, Bhardwaj C, Milasinovic S, Carlson RP, Gordon RJ, Hanley L (2013) Molecular imaging and depth profiling of biomaterials interfaces by femtosecond laser desorption postionization mass spectrometry. ACS Appl Mater Interfaces 5:9269–9275

15. Takahashi LK, Zhou J, Wilson KR, Leone SR, Ahmed M (2009) Imaging with mass spectrometry: a secondary ion and VUV-photoionization study of ion-sputtered atoms and clusters from GaAs and Au. J Phys Chem A 113:4035–4044

16. Sabbah H, Morrow AL, Pomerantz AE, Zare RN (2011) Evidence for island structures as the dominant architecture of asphaltenes. Energ Fuel 25:1597

17. Pan Y, Yin H, Zhang T, Guo H, Sheng L, Qi F (2008) The characterization of selected drugs with infrared laser desorption/tunable synchrotron vacuum ultraviolet photoionization mass spectrometry. Rapid Commun Mass Spectrom 22:2515–2520

18. Elsila JE, de Leon NP, Zare RN (2004) Factors affecting quantitative analysis in laser desorption/laser ionization mass spectrometry. Anal Chem 76:2430–2437

19. Pomerantz AE, Hammond MR, Morrow AL, Mullins OC, Zare RN (2008) Two-step laser mass spectrometry of asphaltenes. J Am Chem Soc 130:7216–7217

20. Getty SA, Brinckerhoff WB, Cornish T, Ecelberger S, Floyd M (2012) Compact two-step laser time-of-flight mass spectrometer for in situ analyses of aromatic organics on planetary missions. Rapid Commun Mass Spectrom 26:2786–2790

21. Misra P, Dubinski MA (eds) (2002) Ultraviolet spectroscopy and VU lasers. Marcel Dekker, New York, NY

Chapter 19

Data Processing and Analysis for Mass Spectrometry Imaging

Jiangjiang Liu, Xingchuang Xiong, and Zheng Ouyang

Abstract

Mass spectrometry imaging produces large numbers of spectra that need to be efficiently stored, processed, and analyzed. In this chapter, we describe the protocol and methods for data processing, visualization, and statistical analysis, with related techniques and tools available presented. Examples are given with data collected for a 3D MS imaging of a mouse brain and 2D MS imaging of human bladder tissues.

Key words Mass spectrometry imaging, Data processing, Tissue imaging, Statistical analysis, Desorption electrospray ionization mass spectrometry

1 Introduction

Mass spectrometry imaging (MSI) has been deployed for a wide range of biological research and applications as it has the capability to provide chemical distributions on samples. MSI enables the collection of information for the chemical distribution that needs to be extracted from the spectra recorded at different locations on a sample surface. This information is then further used in data analysis with the pathological information of the samples for biomarker discovery or clinical diagnosis.

Data processing serves as a bridge between the raw data and data visualization or analysis. Although data processing could be different from method to method due to different samples and practical restrictions, the basic steps of data processing for MS imaging are similar, including the registration of the spectra with the spatial location, transfer of the data format, checking the data quality, selection of the peaks of interest, and storing the extracted spectral information in database for fast and efficient search. Typically a new data set can be constructed with reduced amount of data after the data processing.

Lin He (ed.), *Mass Spectrometry Imaging of Small Molecules*, Methods in Molecular Biology, vol. 1203, DOI 10.1007/978-1-4939-1357-2_19, © Springer Science+Business Media New York 2015

The data visualization produces ion intensity maps for distributions of specific compounds, which can be directly compared with images obtained by traditional H&E staining and other medical imaging techniques, such as magnetic resonance imaging (MRI), computed tomography (CT), and positron emission tomography (PET). The correlations among the distributions can be used for discovering biomarkers or used for clinical diagnosis. Instead of producing the map of ion intensity, maps reflecting statistical analysis can also be generated for the correlations. Due to the extreme complexity of biological samples and strong matrix interferences, multivariate methods toned to be developed and applied, such as principal component analysis (PCA), clustering methods, and factorization methods [1]. These statistic analyses might reveal the difference between normal and disease tissues by comparing signal intensities of single or multiple biomarkers. The developed algorithm can be applied with the identified biomarkers for clinical diagnosis [2–5].

In this chapter, we describe the data processing and analysis using a data set acquired for 36 sections of a mouse brain and human bladder tissues. The data reduction, construction of 3D data set, generation of 2D and 3D images, as well as the data analysis in 2D and 3D data space are introduced.

2   Materials

A wide variety of computing systems can be used for the data processing and data analysis for MS imaging. An example system is a Dell Precision™ workstation T1650 with Intel Core i3 processor, 8 GB DDR3 1,600 MHz memory, 2 TB hard drive, and Windows 7 operating systems (Dell Inc. Round Rock, Texas, US). Microsoft Visual C++ (Microsoft Corporation, Redmond, Washington, USA) and MATLAB (MathWorks Inc., Natick, Massachusetts, USA) are used for programming and running customized software.

The data sets for the 36 sections of the mouse brain [6] and human bladder tissues [7] were recorded using desorption electrospray ionization (DESI) with an LTQ or an Exactive Orbitrap mass spectrometer (Thermo Scientific, San Jose, CA).

3   Methods

Data processing starts with the raw mass spectra collected by MS imaging measurements, which need to be registered with the locations of the sampling points before data reduction and data correlation. The mass spectrometric information is then greatly reduced, normalized, and corrected, ready for data visualization or further statistic analysis [8]. All data used for data processing were collected from mouse brain tissue sections using DESI-MS. All data used for data analysis were collected from human bladder cancerous and adjacent normal tissue sections using DESI-MS.

3.1 Data Acquisition, Storage, and Registration

1. Choose appropriate MS imaging method available to perform the imaging analysis on tissue sections (*see* **Note 1**) [9]. In this chapter, data recorded using DESI-MS are used for examples.

2. Set and optimize the parameters for MS imaging to meet the analytical requirements, e.g., spatial resolution, mass resolution, m/z range, polarities, and MS scan modes. These parameters have impacts on the data size and are used as references for data analysis [10].

3. Record a series of mass spectra by scanning the tissue section pixel by pixel with the ionization source. Correlate the scan number with the physical locations of the sampling points.

4. Save collected spectra as individual files on hard drive (*see* **Note 2**). Use a naming system for the spectral files reflecting the location of the sampling point for each file. A small database or an index file can also be sued for the correlation between the sampling locations and the file names of the spectra.

5. Set the first sampling point at the bottom-left corner of the tissue section as the original point with a coordinate $(0, 0)$. Set each other sampling point with a coordinate (x, y) based on its relative x-y position to the original point $(0, 0)$ (*see* **Note 3**). For 3D MS imaging with multiple tissue sections scanned, add the z coordinate value of each tissue section.

3.2 Data Reduction

1. Deploy peak detection and peak alignment method (PD&PA) on the imaging data set obtained from MS imaging of tissue sections. *See* **Note 4** for the comparison of the PD&PA and the simple bin methods for peak identification.

2. Perform a statistical analysis on all spectra (3,450 spectra in this case) acquired from the entire tissue section (50×69 sampling points in this case). Generate a histogram of the number of peaks at different peak intensity (Fig. 1).

3. Assign the maxima in the histogram as noise level. Set the threshold of signal-to-noise ratio (S/N) as 3. Identify peaks in raw spectra with signal intensity of three times of the noise level and higher as "real peaks." Identify other peaks with signal intensity lower than three times of the noise level as noise or background peaks. Perform a data reduction by removing these identified noise and background peaks (*see* **Note 5**).

4. Perform peak alignment to assign accurate m/z values on "real peaks" to eliminate mass shifts for the same analytes in spectra that are recorded from different sampling points on tissue sections (Fig. 2a). As an example, Fig. 2b and c shows the peak alignment for phosphatidylinositols (PS, 18:0/22:6).

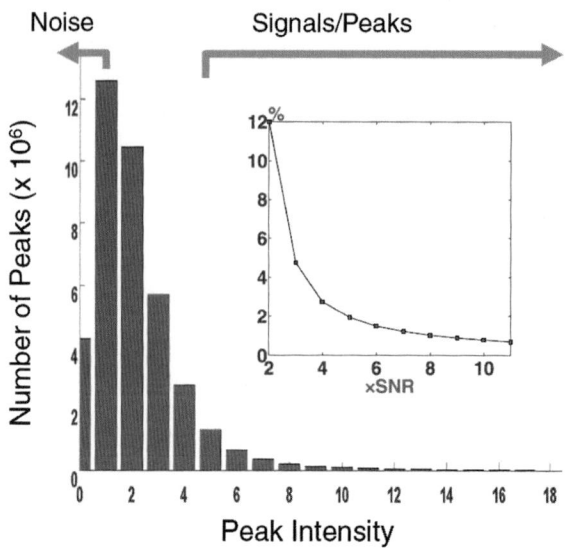

Fig. 1 A histogram of peak intensity based on the statistical analysis of 3,450 spectra recorded from a tissue section of mouse brain. *Inset*, percentage of the original data which can be retained after data reduction as a function of the threshold which is set for identification of "real peaks." Figure taken from reference [8] with permission

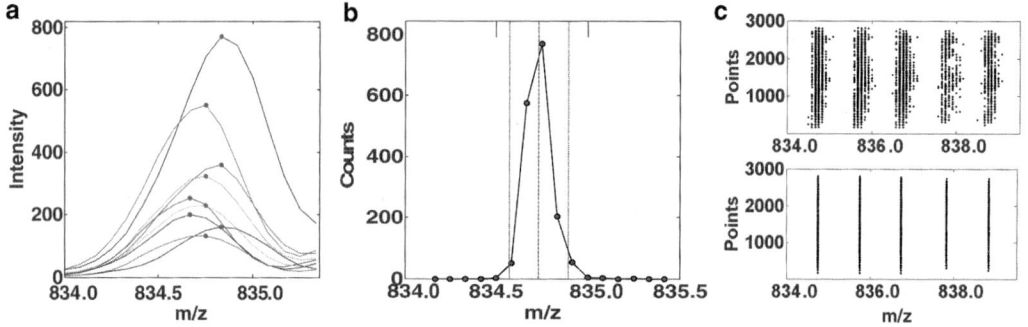

Fig. 2 (**a**) Plotting of peaks (PS 18:0/22:6, *m/z* 834.6) extracted from different spectra, showing mass shifts among different scans. The maximum of each plot is marked with *a red dot*. (**b**) Distribution of *m/z* value for the peaks (PS 18:0/22:6) extracted from all spectra. (**c**) Plotting of the positions of peaks extracted within a mass range *m/z* 834-839 before (*top*) and after peak alignment (*bottom*). Figure taken from reference [8] with permission

(a) Extract the peaks around *m/z* 834.6 in all 3,450 spectra that are obtained from the same tissue section. Set the *m/z* window based on the MS analysis resolution of the mass spectrometer, which is 0.1 Th for the data recorded using an Exactive. Plot the counts of peaks within each *m/z* window as a function of the *m/z* value (Fig. 2b).

(b) Fit the counts of peaks around *m/z* 834.6 to a Gaussian distribution.

(c) Determine the maximum in the Gaussian distribution and assign the corresponding m/z value at this point to all the peaks within the m/z windows in all spectra (Fig. 2c and *see* **Notes 6–8**).

3.3 2D Imaging Data Visualization and Analysis

3.3.1 2D Imaging Visualization

1. Convert the mass spectra files (.raw) collected using LTQ or Exactive with Xcalibur 2.0 into Analyze 7.5 format files (.img, .hdr, and .t2m files) using ImgConverter v3.0 [10] or home-written programs [11]. The number of pixels on x and y dimensions is required for file format conversion. Conversion of raw data files recorded using other mass spectrometers can also be done (*see* **Note 9**).

2. Load the converted files into visualization software, such as Biomap [10], Datacube Explorer [12], and MITICS [13]. Choose a peak for data visualization with appropriate color template, rainbow color scale, and contrast. Save the created image file. Repeat this process to create ion maps of all peaks of interest [10].

3. Overlay two or more ion maps using the overlay function of Biomap to show the distributions of different compounds when necessary (Fig. 3) [10].

3.3.2 2D Imaging Data Analysis

1. Split tissue sections into training set and validation set. Collect spectra from both tissue sets using DESI imaging and obtain the training imaging data set and validation imaging data set, respectively.

2. Resample all pixels in the imaging data sets to unit resolution. Calculate the area under the curve of all spectra. Normalize all spectra using the median area value by scaling each pixel to this value.

3. Deploy principal component analysis (PCA) to the training data set collected from the training tissue sections. Generate a set of principal components (PCs). List the eigenvalues and

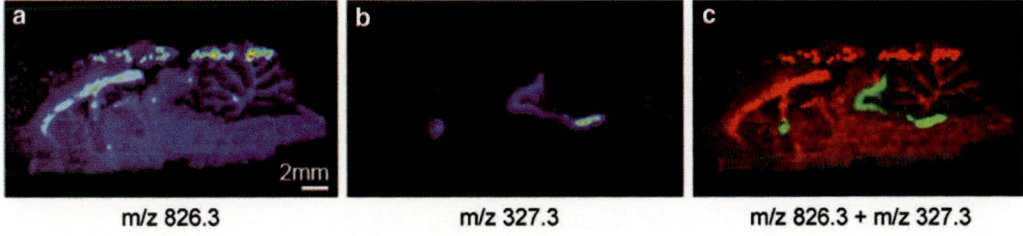

m/z 826.3   m/z 327.3   m/z 826.3 + m/z 327.3

Fig. 3 DESI imaging of a rat brain tissue section. The rat received clozapine and was killed after 45 min. (**a**) Distribution of phosphatidylcholine (PC 36:1), $[M+K]^+$ *m/z* 826.3; (**b**) distribution of clozapine $[M+H]^+$ *m/z* 327.3; (**c**) image of overlaid phosphatidylcholine and clozapine created with Biomap. Figure taken from reference [10] with permission

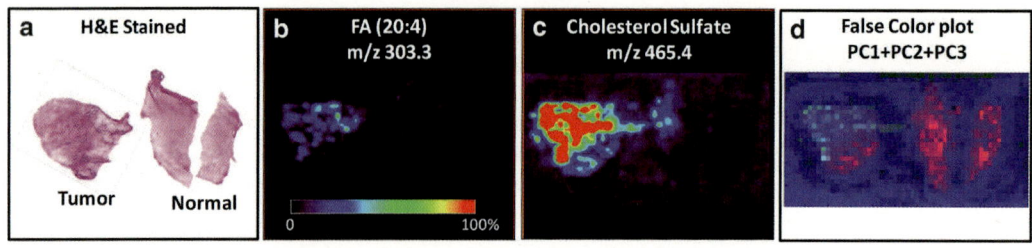

Fig. 4 (**a**) Optical image of the H&E-stained tissue sections including the areas of cancer and adjacent normal tissue. DESI imaging of prostate tissues showing the distributions of (**b**) FA (20:4), *m/z* 303.3, and (**c**) cholesterol sulfate, *m/z* 465.4. (**d**) Image developed with PCA; the false colors plotted here are generated on the basis of PC1, PC2, and PC3. Figure adapted with permission from reference [14]. Copyright 2010 American Chemical Society

eigenvectors in an order of decreasing component variance and pick the first three principal components for data visualization (*see* **Note 10**).

4. Deploy PCA to the validation data set using the PCs obtained from training data set. Scale each set of eigenvalues over the range of 0-255 and assign a color on it within the RGB color space. Eigenvalues of PC1 are converted into the red color channel, eigenvalues of PC2 are converted into the green color channel, and eigenvalues of PC3 are converted into the blue color channel. Assign the RGB color value on each corresponding pixel and create an image based on these pixels for data visualization (Fig. 4) (*see* **Note 11**).

5. Deploy H&E stain pathological analysis on the tissue sections. Correlate the visualized imaging data with H&E stain pathological analysis (*see* **Notes 12** and **13**).

3.4 3D Imaging Data Visualization and Analysis

Additional steps in data processing are required to build a data set in 3D data space.

3.4.1 Tissue Section Alignment

Data analysis can be performed to identify the shapes of the tissue sections and some major features that can be used in the computer-assisted alignment of the tissue sections. This is important for creating a valid 3D data set.

1. Classify MS images into sample and substrate regions with self-organizing feature map (SOFM) artificial neural network method. Identify the tissue sample region by applying SOFM on spectra with original signal intensity to separate the sample region from background (Fig. 5a).

2. For mouse brain tissue 3D imaging, apply SOFM twice on each MS image to differentiate white matter region and gray matter regions (Fig. 5b).

3. Overlay two MS images obtained from the adjacent tissue sections and calculate the counts of mismatched pixels between the two MS images based on the color differences (Fig. 5c and d).

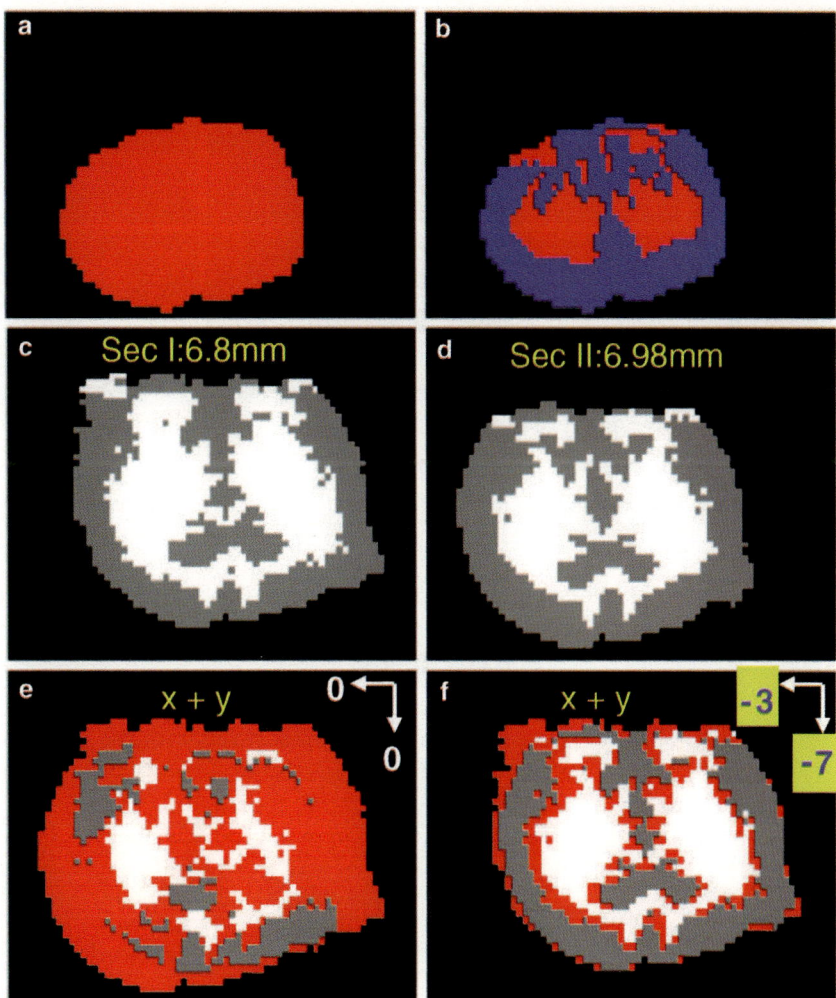

Fig. 5 (**a**) Apply SOFM on the image of a tissue section to obtain a classification into (**a**) two categories, tissue and background, and (**b**) three categories, gray matter, white matter, and background. (**c**, **d**) Images of two adjacent tissue sections (I and II) with three features classified with SOFM. The two images are not aligned. Image can be created by overlapping the sections I and II (**e**) without alignment and (**f**) with alignment. The section II is moved 7 pixels up and 3 pixels to the left. From (**c**) to (**f**), *gray* and *white color* represents *gray matter* and *white matter*, respectively. *Red region* represents the mismatched pixels between sections I and II. Figure taken from reference [8] with permission

4. Align and rotate MS images to minimize the mismatch among different MS images obtained from different tissue sections (Fig. 5e, f and *see* **Note 14**).

3.4.2 Intersection Data Normalization and Image Insertion

1. For 3D imaging, perform interpolation to generate additional layers to improve the smoothness of 3D-visualized images after data set reconstruction (*see* **Note 15**).

2. Generate inserted data for the appropriate image component between two adjacent real layers with linear interpolation

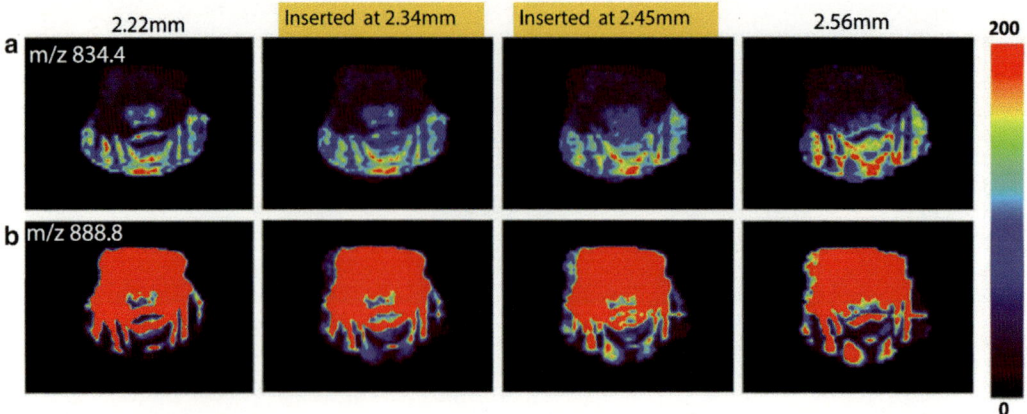

Fig. 6 The 2D images of the distribution of (**a**) PS 18:0/22:6 (*m/z* 834.6) and (**b**) ST 24:1(*m/z* 888.8) generated from the two actual tissue sections (the *most left* and the *most right* images). Figure taken from reference [8] with permission

method. Equation 1 shows how to generate an insert pixel with a coordinate (x_0, y_0) on the insert layer z based on data interpolation of corresponding pixels with the same coordinate (x_0, y_0) on layers z_1 and z_2:

$$P = P_1 + \left(P_2 - P_1\right)\frac{z - z_1}{z_2 - z_1} \tag{1}$$

where P_1 and P_2 are the signal intensities of the MS peaks obtained at the sampling points with the coordinates (x_0, y_0) on layers z_1 and z_2, respectively.

3. Perform 65,550 data interpolations on 19 lipid peaks over 3,450 pixels to generate one insert layer. Construct a 3D data set containing all data obtained from both the original layers and the insert layers. Figure 6 shows the two insert layers with the distributions of PS 18:0/22:6 (*m/z* 834.6) and sulfatides (ST, 24:1, *m/z* 888.8) generated with linear interpolation method (*see* **Note 16**).

3.4.3 3D Imaging Data Visualization

1. Perform 3D visualization on the 3D data set using the 3D Visualization Module in MATLAB (*see* **Note 17**).

2. Generate both 2D and 3D images in three different forms, including iso-surface, slice surface, and subvolume (Fig. 7).

3. Assign appropriate colors to different compounds and use the color intensity to present the relative abundance of this compound in the tissue section. Generate the distribution of different compounds in both 2D and 3D formats.

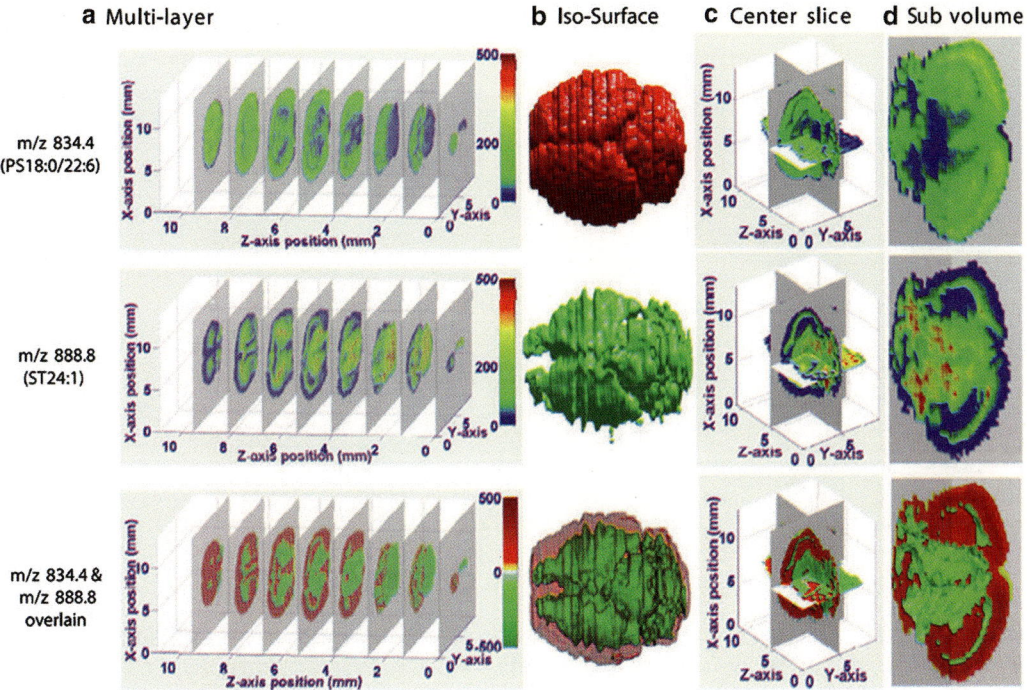

Fig. 7 Visualization of 3D data set collected from tissue sections with MSI. (**a**) 2D images of the distribution of selected compounds present in multiple layers, (**b**) iso-surface views, (**c**) center slice views, and (**d**) subvolume views of the reconstructed 3D imaging of PS 18:0/22:6 (*top*), ST24:1 (*middle*), and both (*bottom*). Figure taken from reference [8] with permission

3.4.4 3D Imaging Statistical Analysis

1. Deploy statistical analysis to the 3D data set which retains the original spectral information. The distribution of compounds in the 3D volume can be visualized for presentation:
   Example: k-mean clustering analysis

   (a) Perform a k-mean clustering analysis to 19 lipids identified in the mouse brain using MATLAB. Classify the mouse brain into two regions based on the results obtained from the k-mean clustering analysis. Figure 8 shows the regions generated by k-mean clustering analysis, which correspond to the white matter and gray matter, respectively.

   (b) Extract the averaged spectra from the two regions (Fig. 8d and e). Identify the dominant peaks at m/z 834.6 (PS 18:0/22:6) in region 1 and m/z 888.8 (ST 24:1) in region.

2. The spectral information retained in the 3D data set can be further processed for biomarker discovery when necessary.

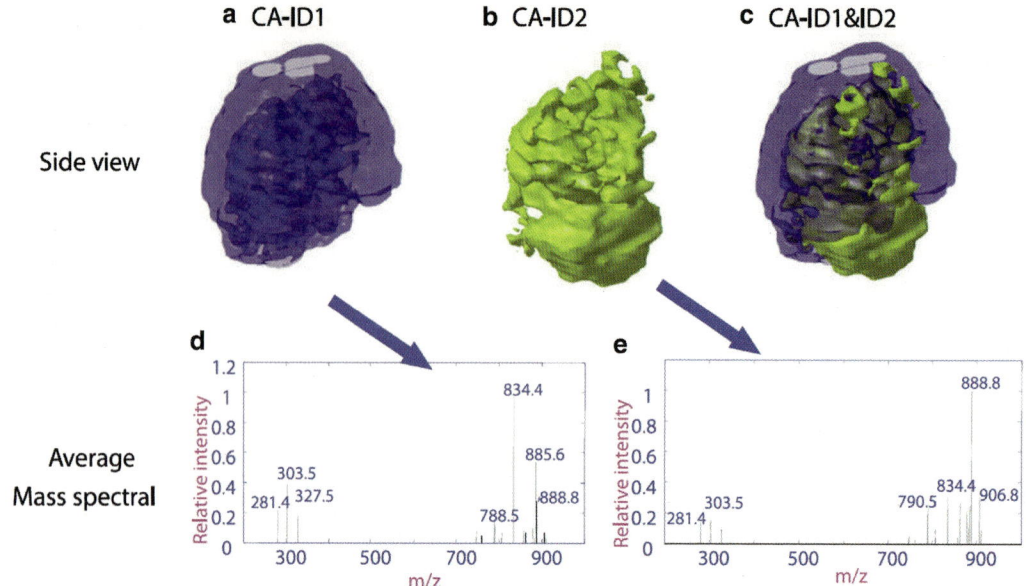

Fig. 8 Deploy a *k*-mean clustering method on the 3D data set and classify the mouse brain into two regions. Side views of (**a**) region 1, (**b**) region 2, and (**c**) the overlap of the two regions. Two averaged mass spectra are extracted from regions (**d**) 1 and (**e**) 2. Figure taken from reference [8] with permission

4   Notes

1. A general guideline: secondary ion mass spectrometry (SIMS) for MSI of small molecules and metals; matrix-assisted laser desorption/ionization (MALDI) for MSI of large biomolecules; desorption electrospray ionization (DESI) for MSI of small molecules with moderate and high polarity.

2. Spectra can be collected with different MSI setup, e.g., mass spectrometers, MS operational interfaces, and software. The stored spectrum files can be opened with certain software depending on the instruments and software used for data collection. imzML is a promising data format developed for data storage, exchange, and processing among different instrumentations [15].

3. The *x* and *y* positions of the sampling point corresponding to each spectrum can be determined by the scanning speed and step length of the moving stage used for moving tissue section in the *x* and *y* directions, respectively.

4. Figure 9 shows the peak identification process on the same spectral profile using PD&PA method and bin method, respectively. Both methods can identify several peaks from raw spectrum, but only the PD&PA method reserves the resolution of raw spectrum and provides accurate *m/z* value of these peaks.

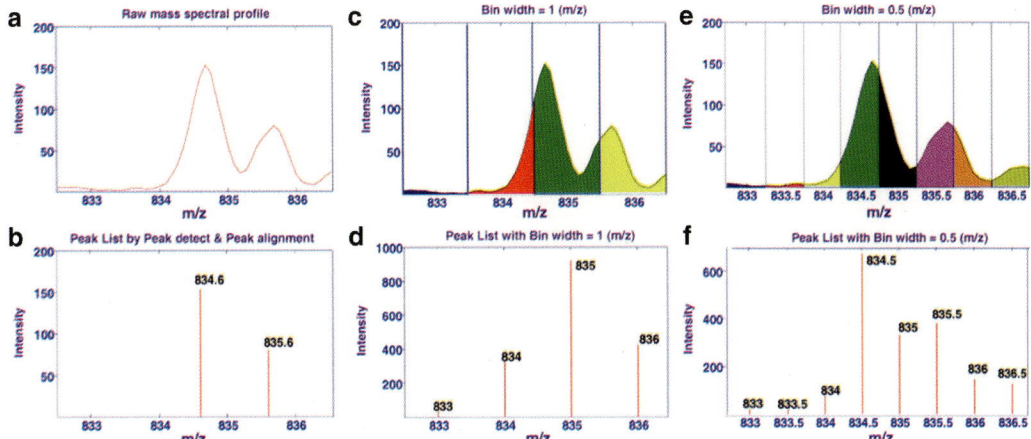

Fig. 9 Comparison of PD&PA (**a**, **b**) and bin (**c–f**) methods for peak identification. Figure taken from reference [8] with permission

5. Setting a higher S/N for peak detection leads to a further reduction on the data set but this can only be deployed with a confidence based on prior knowledge of the tissue section. On the other hand, setting a lower S/N for peak detection would help saving the significant peaks of low abundance but the data involved in the latter stage of analysis would be significantly increased.

6. It is not necessary to perform the statistical analysis if internal reference peaks are available for mass calibration of each spectrum. But the peak positions in different spectra still need to be aligned as shown in Fig. 2c.

7. If two or more local maxima are observed after the Gaussian fitting, the mass accuracy and resolution of the mass spectrometer then need to be considered to determine if it is appreciate to assign multiple peaks.

8. Peak binning is another popular data reduction method used for MS imaging. A comparison of the bin method and PD&PA method used here is shown in Table 1. The method used here provides a better reduction on the data set while still retaining the accurate mass information [8].

9. MS imaging data can be generated with Biomap, Datacube Explorer, MITICS, SpectViewer, Mirion, OpenMS, and MapQuant. A detailed list of currently available software for mass spectrometry imaging can be found in Table 2 [13]. Some other open-source data processing packages are also available for 3D visualization, such as Visualization ToolKit (http://www.vtk.org) and vol3d (http://www.mathworks.com/matlabcentral/fileexchange/22940-vol3d-v2).

Table 1
A comparison of data reduction on data sets obtained with LTQ and Orbitrap using binning method and PD&PA method, respectively

Instrument	*m/z* Range	Pixels per tissue section	Raw data size	Strategy	*m/z* Window	Ion maps	Data size
LTQ	*m/z* 150–1,100	50×69	153 MB	Bin	1 Th	950	12.5 MB
				PD&PA	0.1 Th	80	0.98 MB
Orbitrap	*m/z* 780–920	36×123	265 MB	Bin	0.1 Th	1400	23.6 MB
				PD&PA	0.01 Th	84	1.33 MB

10. Besides PCA, many statistical analysis methods are available for data analysis for MSI: (1) clustering methods, including hierarchical clustering, *k*-means clustering, and fuzzy c-means; (2) factorization methods, including nonnegative matrix factorization (NNMF), probabilistic latent semantic analysis (PLSA), and maximum autocorrelation factor analysis (MAF) [1, 16, 17].

11. An interactive hyperspectral approach was developed to explore and interpret DESI imaging results for differentiation of cancerous and normal tissue [18].

12. DESI imaging, MALDI imaging, and H&E pathological analysis can be performed on the same sections in serial [19].

13. DESI imaging has been deployed for tissue profiling and clinical diagnosis on a wide range of animal and human tissues, including seminoma [3], brain [2], bladder [5], spinal cord [20, 21], kidney [4], and adrenal gland [22].

14. The sample tissue region identified with SFOM sometimes needs to be reshaped in some cases due to the stretching or shrinking of the tissue section.

15. Based on the assumption that the distributions of biological molecules inside tissues are continuous.

16. Figure 10 shows the comparison of generation of insert data with different interpolation methods, including nearest, linear, and cubic-spline-interpolation methods. As an example, data insertions are performed for PS 18:0/22:6 (*m/z* 834.6) and sulfatides (ST, 24:1, *m/z* 888.8) using the three interpolation methods. Typically, there is no significance among the insert images generated with these interpolation methods [8].

17. 3D imaging visualization was performed on 36 tissue sections. 364 insert layer images were generated from the MS images of the 36 tissue sections. The total 400 layers with 26.6 μm interval are used for the reconstruction of 3D data set.

Table 2
Software available for data processing of mass spectrometry imaging

Software	Function	Input format	Output format	Spectrometer compatibility	Availability	Free
MMSIT	Acquisition	Instrument format .dat	Analyze 7.5	AB MALDI/ TOF	www.maldi-si.org	Yes
4700/4800 Imaging	Acquisition	Instrument format	Analyze 7.5	AB MALDI- TOF/TOF	www.maldi-msi.org	Yes
Create Target Analyze This!	Acquisition Data conversion	Binary	Analyze 7.5	BD MALDI/ TOF MALDI- TOF/TOF	www.maldi-msi.org	Yes
Axima2 Analyze	Acquisition Data conversion	Instrument format	Analyze 7.5	SZU	www.maldi-msi.org	Yes
oMALDI Server 5.1	Acquisition Image reconstruction	Unknown	Analyze 7.5	AB/MDS SCIEX oMALDI QSTAR Hybrid LC/ MS/MS	AB	No
Flex Imaging	Acquisition Image reconstruction	Instrument format .img	.bmp .jpeg,.jpg .tif,.tiff	BD MALDI/ TOF MALDI- TOF/TOF	BD	No
Tissue view 1.0	Acquisition Image reconstruction	Instrument format	.tiff .jpeg Analyze 7.5	AB/MDS SCIEX MALDI- TOF/TOF QSTAR	AB	No
Image Quest	Acquisition Data conversion	Instrument format	Unknown	TS MALDI LTQ XL LTQ Orbitrap hybrid series	TS	No
BIOMAP	Image reconstruction	Analyze 7.5 .img .tiff .dicom .pnp	.tiff .jpeg Analyze 7.5	AB MALDI- TOF 4700 4800 MALDI- TOF/TOF MALDI Micro MX(Waters)	www.maldi-msi.org	Yes
MITICS	Acquisition Image reconstruction	.xml	.bmp .jpg .jp2 .gif .pict .png .tif	AB MALDI/ TOF BD MALDI/ TOF TOF/ TOF	Isabelle.fournier@ univ-lille1.fr (I. Fournier), michel.salzet@ univ-lille1.fr (M. Salzet).	Yes

AB Applied Biosystems, *BD* Bruker Daltonics, *SZU* Shimadzu, *TS* Thermo Scientific

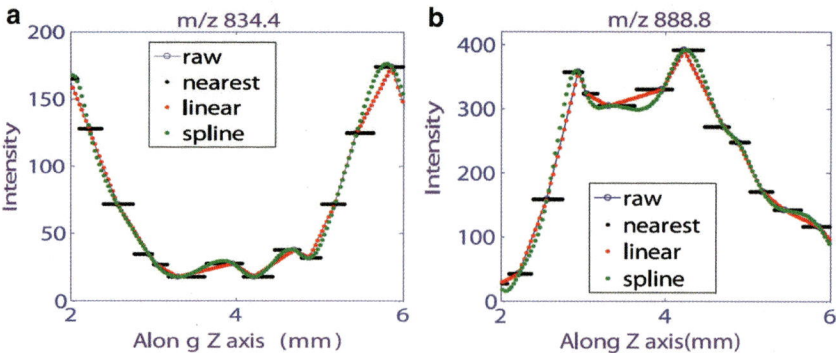

Fig. 10 The intensities of (**a**) PS 18:0/22:6 and (**b**) ST 24:1 in different spectra collected from a series of tissue sections and the trend lines for data insertion generated with different interpolation methods. Figure taken from reference [8] with permission

References

1. Jones EA, Deininger SO, Hogendoorn PCW, Deelder AM, McDonnell LA (2012) Imaging mass spectrometry statistical analysis. J Proteomics 75(16):4962–4989

2. Eberlin LS, Norton I, Dill AL, Golby AJ, Ligon KL, Santagata S, Cooks RG, Agar NYR (2012) Classifying human brain tumors by lipid imaging with mass spectrometry. Cancer Res 72(3):645–654

3. Masterson TA, Dill AL, Eberlin LS, Mattarozzi M, Cheng L, Beck SDW, Bianchi F, Cooks RG (2011) Distinctive glycerophospholipid profiles of human seminoma and adjacent normal tissues by desorption electrospray ionization imaging mass spectrometry. J Am Soc Mass Spectrom 22(8):1326–1333

4. Dill AL, Eberlin LS, Zheng C, Costa AB, Ifa DR, Cheng LA, Masterson TA, Koch MO, Vitek O, Cooks RG (2010) Multivariate statistical differentiation of renal cell carcinomas based on lipidomic analysis by ambient ionization imaging mass spectrometry. Anal Bioanal Chem 398(7–8):2969–2978

5. Dill AL, Eberlin LS, Costa AB, Zheng C, Ifa DR, Cheng LA, Masterson TA, Koch MO, Vitek O, Cooks RG (2011) Multivariate statistical identification of human bladder carcinomas using ambient ionization imaging mass spectrometry. Chemistry 17(10):2897–2902

6. Eberlin LS, Ifa DR, Wu C, Cooks RG (2010) Three-dimensional visualization of mouse brain by lipid analysis using ambient ionization mass spectrometry. Angew Chem Int Ed Engl 49(5):873–876

7. Cooks RG, Manicke NE, Dill AL, Ifa DR, Eberlin LS, Costa AB, Wang H, Huang GM, Zheng OY (2011) New ionization methods and miniature mass spectrometers for biomedicine: DESI imaging for cancer diagnostics and paper spray ionization for therapeutic drug monitoring. Faraday Discuss 149:247–267

8. Xiong XC, Xu W, Eberlin LS, Wiseman JM, Fang X, Jiang Y, Huang ZJ, Zhang YK, Cooks RG, Ouyang Z (2012) Data processing for 3D mass spectrometry imaging. J Am Soc Mass Spectrom 23(6):1147–1156

9. Liu JJ, Ouyang Z (2013) Mass spectrometry imaging for biomedical applications. Anal Bioanal Chem 405(17):5645–5653

10. Wiseman JM, Ifa DR, Venter A, Cooks RG (2008) Ambient molecular imaging by desorption electrospray ionization mass spectrometry. Nat Protoc 3(3):517–524

11. Robichaud G, Garrard KP, Barry JA, Muddiman DC (2013) MSiReader: an open-source interface to view and analyze high resolving power MS imaging files on Matlab platform. J Am Soc Mass Spectrom 24(5):718–721

12. Smith DF, Kharchenko A, Konijnenburg M, Klinkert I, Pasa-Tolic L, Heeren RMA (2012) Advanced mass calibration and visualization for FT-ICR mass spectrometry imaging. J Am Soc Mass Spectrom 23(11):1865–1872

13. Jardin-Mathe O, Bonnel D, Franck J, Wisztorski M, Macagno E, Fournier I, Salzet M (2008) MITICS (MALDI Imaging Team Imaging Computing System): a new open source mass spectrometry imaging software. J Proteomics 71(3):332–345

14. Eberlin LS, Dill AL, Costa AB, Ifa DR, Cheng L, Masterson T, Koch M, Ratliff TL, Cooks RG (2010) Cholesterol sulfate imaging in human prostate cancer tissue by desorption

electrospray ionization mass spectrometry. Anal Chem 82(9):3430–3434

15. Schramm T, Hester A, Klinkert I, Both JP, Heeren RMA, Brunelle A, Laprevote O, Desbenoit N, Robbe MF, Stoeckli M, Spengler B, Rompp A (2012) imzML—a common data format for the flexible exchange and processing of mass spectrometry imaging data. J Proteomics 75(16):5106–5110

16. Thomas A, Patterson NH, Marcinkiewicz MM, Lazaris A, Metrakos P, Chaurand P (2013) Histology-driven data mining of lipid signatures from multiple imaging mass spectrometry analyses: application to human colorectal cancer liver metastasis biopsies. Anal Chem 85(5): 2860–2866

17. Fonville JM, Carter CL, Pizarro L, Steven RT, Palmer AD, Griffiths RL, Lalor PF, Lindon JC, Nicholson JK, Holmes E, Bunch J (2013) Hyperspectral visualization of mass spectrometry imaging data. Anal Chem 85(3):1415–1423

18. Pirro V, Eberlin LS, Oliveri P, Cooks RG (2012) Interactive hyperspectral approach for exploring and interpreting DESI-MS images of cancerous and normal tissue sections. Analyst 137(10):2374–2380

19. Eberlin LS, Liu XH, Ferreira CR, Santagata S, Agar NYR, Cooks RG (2011) Desorption electrospray ionization then MALDI mass spectrometry imaging of lipid and protein distributions in single tissue sections. Anal Chem 83(22):8366–8371

20. Girod M, Shi YZ, Cheng JX, Cooks RG (2010) Desorption electrospray ionization imaging mass spectrometry of lipids in rat spinal cord. J Am Soc Mass Spectrom 21(7): 1177–1189

21. Girod M, Shi YZ, Cheng JX, Cooks RG (2011) Mapping lipid alterations in traumatically injured rat spinal cord by desorption electrospray ionization imaging mass spectrometry. Anal Chem 83(1):207–215

22. Wu CP, Ifa DR, Manicke NE, Cooks RG (2010) Molecular imaging of adrenal gland by desorption electrospray ionization mass spectrometry. Analyst 135(1):28–32

INDEX

Lin He (ed.), *Mass Spectrometry Imaging of Small Molecules*, Methods in Molecular Biology, vol. 1203,
DOI 10.1007/978-1-4939-1357-2, © Springer Science+Business Media New York 2015

Printed by Printforce, the Netherlands